CURTISS SOC/SON SEAGULL FLEET SCOUT/OBSERVATION AIRCRAFT

Above, Seagull prototype XOSC-1 BuNo 9413 in flight with original open cockpits and amphibious gear on the main float. Production SOCs had a full NACA cowl, enclosed cockpit, and flight control improvements through ailerons, flaps and slats. The SOCs were not amphibious or carrier capable. (USN)

INTRODUCTION: The catalyst for the 1933 VSO design competition was provided by Alexander deSeversky who made an unsolicited impromptu demonstration to the Navy in 1933 of his new all-metal SEV-3 amphibian. The high performance monoplane was capable of speeds in excess of 200 mph. Although impressed, the Navy did not order a prototype but instead issued design specifications for a new VSO aircraft which called for a fabric wing biplane. Seversky declined to participate in the competition even after the fabric wing restriction was dropped.

In 1933 the Navy solicited the industry for a replacement of the Vought O2U AND O3U floatplanes. From the respondents, three manufacturers, Douglas, Vought and Curtiss, were awarded contracts for prototypes. The Curtiss prototype, the XOSC-1 (contract# 31827), was ordered on 19 June 1933 as was the Douglas XO2D-1 and the Vought XO5U-1. In April 1934 the Douglas entry was the first to be delivered followed by the Vought and Curtiss entries in May 1934. After trials at NAS Anacostia, the Curtiss entry was declared the winner due in a great deal to its improved performance which resulted from the design's use of full-span automatic slots and hand operated flaps on the upper wing.

PERFORMANCE BENEFITS OF THE SOC SLOT AND FLAP SYSTEM: The lift of an ordinary airfoil is approximately proportional to the angle of attack up to the point where the flow breaks down on the upper surface of the wing, resulting in the well-known stall as seen in Figure 1. This breakdown of flow, or stall, occurs with most airfoils (pre-WWII) at approximately 15° angle of attack, which corresponds generally to the attitude of the airplane in a three-point landing.

1930's wind tunnel research and flight testing showed that if a smaller airfoil was placed in the proper position with respect to the airplane wing, so that a slot is formed between this small airfoil and the wing, a fast flowing sheet of air can be run over the top surface of the main wing which will delay the stall to a much higher angle 23° and, as the lift is still proportional to the angle, to a much higher lift as seen in Figure 2 (next page).

The slot's small airfoil, while providing an excellent effect on the stall, had the disadvantage that it added considerably to the drag of the airplane in the level flight attitude and, therefore, would cut down the high speed performance. Further research showed that if this slot was placed on

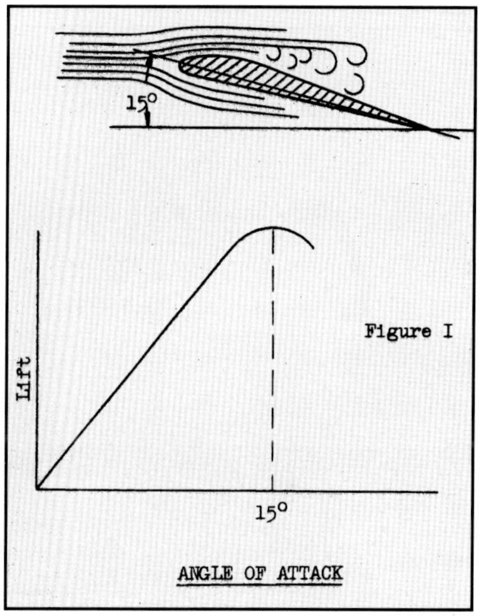

Figure I

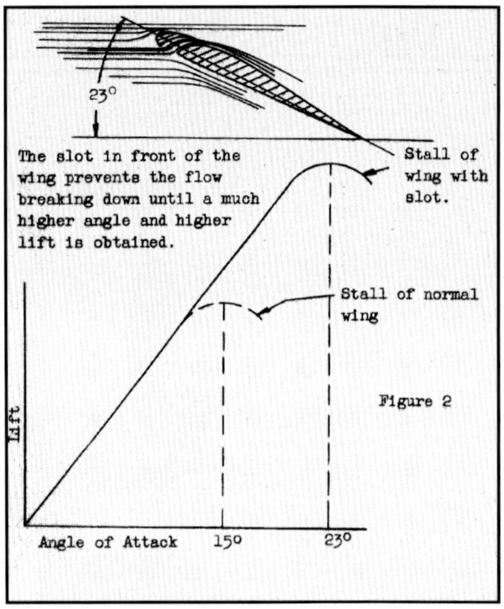

Top-to-bottom: Vought O2U-3 BuNo 8209 at NAS Norfolk while assigned to the Naval Torpedo Experimental Unit in October 1934. (Jim Hawkins) Vought O3U-3 BuNo 9298 at Floyd Bennett Field, NY, in May 1933, (Jim Hawkins) The unsolicited Seversky SEV-1 (X-2106) that was flown to Anacostia at company expense in hopes of securing a high speed scout for cruiser contract. (via Peter Bowers) Looking much like a production SOC, the reworked XO3C-1 BuNo 9413 is seen on 23 April 1935 with enclosed cockpit. (USN)

a supporting track of proper shape and fitted with rollers, the air forces were such that it would operate automatically, extending to the high lift position as the stall is approached and nesting into the main wing at high speed.

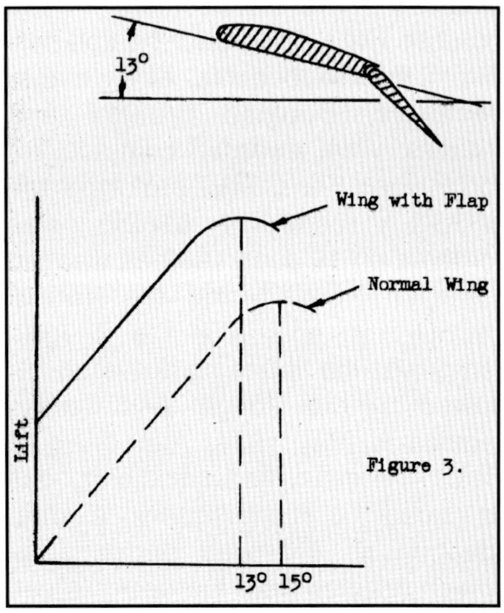

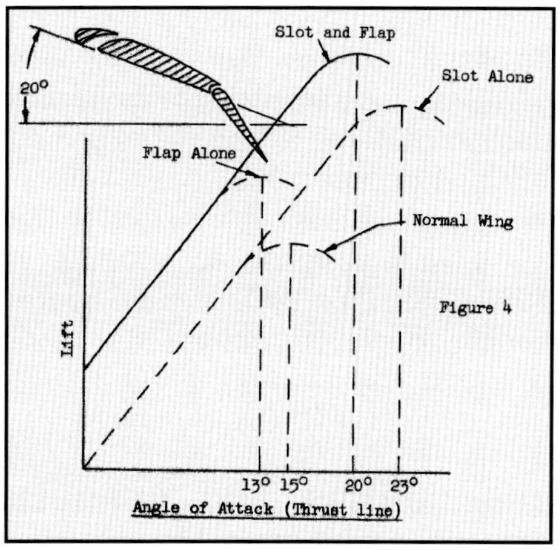

Unfortunately, the full value of the high lift automatic slot could not be obtained unless the airplane was placed at a very high angle of attack, much higher than it is practical to obtain in a normal landing attitude. However, by combining the automatic slot with a trailing edge flap it was possible to obtain the full advantages of both slots and flaps on an airplane having normal controls. The trailing edge flap on an airplane wing has the effect of not only increasing the maximum lift but of shifting the lift curve into the lower angles of attack as seen in Figure 3. With the flaps, maximum lift occurs at approximately 13° angle of attack, which is less than the attitude of the airplane as it approaches the ground for a three-point landing. If both slots and flaps were used on the main wing the full benefits of both devices are obtained. Increases of high lift are retained and the peak of the lift, or the stalling, occurs at slightly above the normal approach angle of the airplane (15°) as seen in Figure 4.

On the SOC, the combination automatic slot and wing flap configuration increased lift approximately 42% which allowed a landing speed of 54 mph without the large increase in wing area which would otherwise be necessary.

On the SOC the slots were fully closed at about 120 knots and could then be locked closed by the pilot if desired. Furthermore, the flaps were restricted to operations below 105 knots and when lowered automatically unlocked the slats.

DOUGLAS PROTOTYPE TEST HISTORY: The Douglas XO2D-1, BuNo 9412, first flew at Clover Field, Santa Monica, CA, on 27 March 1934. It was piloted by Edmund T. Allen and exhibited good flight characteristics. However, the brakes and tail wheel were a major concern. Repeatedly the brakes locked up during taxiing causing the tail wheel to rise up some two feet before rebounding on the runway. The result was structural failure of the aft end of the main float. The problem was finally fixed on 11 April, fifteen flights later, after which it was shipped to NAS Anacostia where it arrived on 23 April.

Three weeks later, an engine change was required after the number three cylinder head burned through. Two days after repairs were made, the right landing gear shock stop failed and the landing gear could not be retracted nor landed on. An emergency water landing was made and the dangling right gear caused the aircraft to nose over. After salvage, the aircraft was stripped of fabric and washed in fresh water before going to Brewster for repairs. The XO2D-1 was returned to flight status on 7 July and continued Navy evaluation until March 1935. During final Navy, trials the amphibious float was replaced by a non-amphibious float as the new Navy specifications dictated.

The XO2D-1 was returned to Douglas for rework where general corrosion was discovered and an overhaul was conducted. After the SOC was awarded the contract, it languished at Douglas until 13 September 1935 when it was trucked to NAS San Diego. The engine and instruments were removed and the XO2D-1 was stricken on 31 October. The airframe was then shipped to the Naval Aircraft Factory where it arrived in December 1935 for use in structural tests.

VOUGHT PROTOTYPE TEST HISTORY: The Vought XO5U-1 (E105), BuNo 9399, first flew at East Hartford, CT, on 8 May 1934. It was piloted by Vought test pilot Paul S. Baker. It was delivered to the Navy under contract C-31130 at NAS Anacostia in May 1934.

The aircraft originally featured detached ailerons located above the upper wing along with conventional ailerons on the lower wing. The final configuration had conventional ailerons and large flaps on the upper wing.

The Navy tested the XO5U-1 at Anacostia and at Norfolk. It was later assigned to VX-3D4 at the Naval Aircraft Factory where it crashed during a water landing in May 1938. During landing the wings sheared off and the aircraft sank. After salvage the XO5U-1 was stricken.

CURTISS PROTOTYPE TEST HISTORY: The XO3C-1, BuNo 9413, was delivered to NAS Anacostia on 4 May 1934.

In March 1935, the aircraft was modified to test all SOC production

At right, the Douglas entry, the XO2D-1 BuNo 9412 on its amphibious gear in 1934. (USN)

aircraft new specifications including enclosed cockpits. As such, it was outwardly indistinguishable from production SOC-1s. It was redesignated XSOC-1 in June 1935.

It was assigned to VX-3D4 in 1938-1939. In 1940, it was operationally deployed aboard the USS Noa and was stricken in June 1941 after a crash at NAS Jacksonville, FL.

THE CURTISS SOC SEAGULL Model 71A: The original specification called for folding wings, amphibious floats and a carrier tail hook. However, the production contract, 41179 dated 23 March 1935, for the Curtiss SOC-1 only called for the folding wing feature and the amphibious and tailhook requirement were deleted. The new specification also incorporated the enclosed cockpit only found on the Douglas entry.

The 135 SOC-1s, BuNos 9856 through 9990, were built to the Bureau of Aeronautics Detail Specification SD-230, dated 25 March 1935. All aircraft were completed at the Buffalo Kenmore facility and delivered with fixed landing gear. Float gear was then installed by the Navy and used interchangeably with the fixed landing gear depending on whether the ship was at sea or in port. The prototype design had been classed as battleship observation but that had been broadened to scout observation for the SOC-1 production aircraft. This change caused the prototype to be redesignated XSOC-1 on 23 March 1935.

The XOSC-1, XO2D-1 and XO5U-1 were powered by a Wright R-1340-12 engine of 550 hp while the SOC-1 was powered by a -18 engine of the same horsepower. Like the pro-

---

Top-to-bottom: the Douglas entry, the XO2D-1 BuNo 9412 on its amphibious gear in 1934. (USN) The XO2D-1 with non-amphibious float and removed tailhook in 1935. Two views of the Vought entry, the XO5U-1 BuNo 9399, with amphibious gear and tailhook and later during water operations. (USN)

totype, the SOC was armed with two .30 caliber Browning M-2 machine guns, gun camera and the Mark III, Mod. 2 telescope sight. One machine gun was mounted in a fixed position on the starboard side of the fuselage and was fired through the propeller arc by the pilot. The second machine gun was flex-mounted in the observer's cockpit. The aft fuselage was fitted with a collapsible turtledeck that was lowered to allow the rear-seat observer/gunner a clear field of fire. Provisions were included for a Mark XLI or Mark XXI Mod. 1 bomb rack under each wing. Other equipment featured a Very pistol and ammunition, eight aircraft float lights, two H.C. smoke grenades, two aircraft parachute flares, pressure fire extinguisher, hand-held fire extinguisher, four-point thermocouple, engine driven generator, one type A battery, GP-1 complete transmitting and receiving radio, chart board, and two-man life raft. Deliveries began on 12 November 1935.

SOC-2 Model 71B: The second version of the SOC was virtually identical to the SOC-1 with only minor changes and a R-1340-22 engine. The 40 aircraft were assigned BuNos 0386 through 0425.

XSO2C-1 Model 71C: A single XSO2C-1 BuNo 0950 was built as an improved version of the SOC. The aircraft featured ailerons on both the top and bottom wings. It was delivered to NAS Anacostia in February 1937. After a crash at Anacostia on 29 June 1941 it was stricken and delivered to the mechanics school at NAS Jacksonville, FL.

SOC-3 Model 71E: Eighty-three SOC-3s, BuNos 1064 through 1146, were built with R-1340-22 engines and minor interior arrangement changes.

At right top-to-bottom: SOC Seagull replacements, the Vought OS2U Kingfisher, Curtiss SO3C Seagull/Seamew, Curtiss SC-1 Seahawk, and the EDO XTE, which did not receive a production contract. (USN)

CURTISS XOS3C-1 PROTOTYPE SCOUT AIRCRAFT

At top left, the XO3C-1 during service evaluations at NAS Anacostia. (USN) Four views of the open cockpit XO3C-1 BuNo 9413 at the Curtiss factory on 3 March 1934. (USN) Below, XO3C-1 on 23 April 1935 with enclosed cockpit added. It was redesignated XSOC-1 in June 1935. (USN)

CURTISS SOC-1 WALKAROUND

At left and below, the first SOC-1 BuNo 9856 on 14 August 1935. Outwardly, there was no visible difference between the SOC-1/-2/-3 and SON-1 except the SOC-1's engine cowl gills. (USN) Above, the first SOC-1 BuNo 9856 on 10 September 1935. (USN) Bottom, the first SOC while assigned to NAS Anacostia for extended testing. (USN)

CURTISS SOC/SON WINGS FOLDED

Above, first SOC-1, BuNo 9856, undergoes floatplane flight testing at NAS Anacostia. (USN) Below, tight storage of an SOC half-way in the hangar of the USS Minneapolis (CA-36) with wings folded on 17 August 1940. (USN)

Above, SON-1 BuNo 1147 during its second landing during rough water tests off Hampton Roads on 15 December 1938. (USN) Below, tight storage of an SOC in the hangar on the Northampton. Note spare wings and wing tip floats hanging from the overhead. (USN)

At top, factory SOC flightline in 1935. (via Peter M. Bowers) Above, first SOC-1 in flight. (USN) Below, good underside view of an SOC-1 landing at NAS Oakland, CA, in September 1940. Note extended slots on upper wing. (Peter M. Bowers)

ONE - OFF CURTISS XS02C-1 SCOUT/OBSERVATION AIRCRAFT

XSO2C-1 BuNo 0950 on 23 January 1937. The aircraft featured ailerons on both the top and bottom wings. It was delivered to NAS Anacostia in February 1937. After a crash at Anacostia on 29 June 1941, it was stricken and delivered to the mechanics school at NAS Jacksonville, FL. (USN) At right, bottom wing flaps extended. Below, XSO2C-1 in flight during testing at NAS Anacostia on 23 March 1937. (USN)

At left, SOC-2 BuNo 0417 assigned to NAS Anacostia on 20 April 1938. (USN) Bottom left, SOC-1 BuNo 9875 assigned to the USS Detroit (CL-8) in May 1936 at Grand Central Airport, Glendale, CA. The Detroit was one of six cruisers that had SOC ship units. These aircraft were not assigned to scouting squadrons and did not carry tail and wing markings. (Willian Swisher)

Naval Aircraft Factory SON-1: An additional forty-four SOC-3s were built at the Naval Aircraft Factory as SON-1s. These were BuNos 1147 through 1190. At the time, the NAF was required by the Vinson-Trammell mandate to manufacture a small quantity of each manufacturer's design to establish cost guidelines. The NAF was originally scheduled to produce the SOC-2 under the SON-1 contract. Production was ordered on 10 June 1937, but in July it was decided to switch to the SOC-3. The NAF designed-in some detail changes on the SON-1. These were modified wing tip floats, a redesigned radio installation, and changes to the covering of the horizontal stabilizer. Production start was delayed some three months due to inaccurate blueprints that did not match actual production parts. The first SON-1, BuNo 1147, first flew on 15 September 1938 in its floatplane configuration. During tests at Anacostia the upper wing fabric came loose and the aircraft was returned to the factory for fixes. After its return to Anacostia, some alarming spin characteristics were discovered which were corrected by changes made to the rudder trim tab. Production deliveries commenced in early 1939 and were completed in late August. After delivery, some of the fixed pitch propellers developed cracks, and in April adjustable pitch propellers were substituted.

SOC-1A, SON-1A and SOC-3A CARRIER CAPABLE SEAGULLS

An unknown number of Seagulls were converted with tailhooks starting in late 1941 for use aboard the Navy's new escort carriers. Included in this group were the three Coast Guard SOC-4s. The Navy took control of these aircraft in 1942 and reassigned them BuNos 48243-48245 before converting them to SOC-3As. The SOC-3As were also modified to carry a centerline depth bomb.

Brief SOC/SON service history: From 1937 through 1940, the Seagull dominated scout/observation units in the fleet. However, by 7 December 1941 almost all SOCs on battleships had been replaced with the fixed wing Vought OS2U Kingfisher.

In 1938, a contract to replace the SOC with the Ranger-powered SO3C was acquired by Curtiss (see Naval Fighters Number 47 Curtiss SO3C Seamew). A protracted development program and numerous problems delayed production until 4 March 1942 when the first production aircraft flew. They joined the fleet in July aboard the USS Cleveland and proved to be a complete failure. They were so underpowered that they could not take off with a full fuel load. They were then withdrawn after an average of two-months service aboard the ships they were assigned to. This withdrawal from service resulted in the SOCs being removed from storage, refurbished, and returned to the fleet in the VCS squadrons. The SOCs were not retired until the war ended.

USCG CURTISS SOC-4

Three factory photos of SOC-4 V171 on 4 March 1938. (USN)

The three Coast Guard SOC-4s received serial numbers V171 through V173 which were painted in black on the aft fuselage under "COAST GUARD". The aircraft were aluminum overall except for the upper wing which was chrome yellow (orange/yellow). The top third of the rudder was dark blue with a white SOC-4 appearing on each side. The vertical rudder stripes were red and white.

At top and right, USCG SOC-4 serial number V171. (Peter M. Bowers collection) Below, V171 taxiing at Miami. (Peter M. Bowers collection)

USCG SOC-4

At right, ski gear for the SOC-4 included in the SOC-4 contract. (National Archives)

Below, USCG SOC-4 serial number V173 in flight from USCG Station Port Angeles, WA. Note USCG serial number V173 is repeated at the bottom of the main float. (Gordon S. Williams)

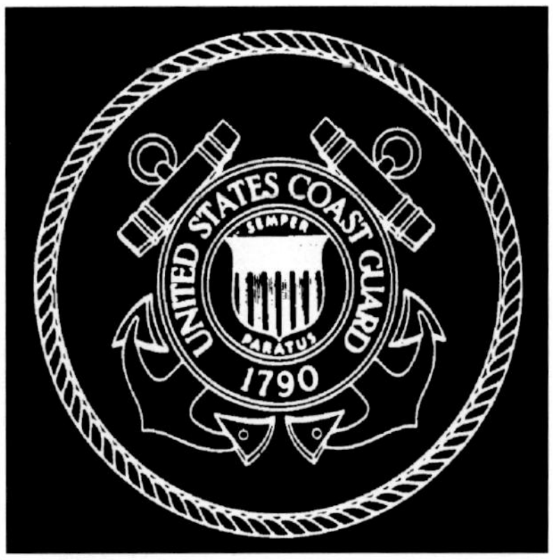

USCG SOC-4 serial number V173 at USCG Station Port Angeles, WA. (Peter M. Bowers collection)

CURTISS SOC/SON AIRCRAFT MARKINGS 1935-1947

When the SOC entered service in 1935, aircraft were finished in overall aluminum with the upper surface of the top wing and 5° of the lower leading edge of the top wing painted chrome yellow (orange yellow). Propeller blades were painted silver and the forward side of the propeller tips were painted in red, yellow, and dark blue 4-inch bands. The corresponding aft side of the propeller tips were painted flat black.

Observation and scout aircraft were further identified with the following tail markings. Battleship Observation Squadrons had separate tail colors for each squadron which covered the complete vertical and horizontal tail. This practice was ordered discontinued on 26 February 1941.

VO-1B/VO-1	**Insignia Red**
VO-2B/VO-2	**White**
VO-3B/VO-3	**True Blue**
VO-4B/VO-4	**Black**
VO-5B/VO-5	**Lemon Yellow***

*** authorized October 1939**

From 1935 through 1 July 1937, Cruiser Scout Squadrons were identified with one wide stripe across the vertical fin and horizontal stabilizer in the following colors:

VS-5B	**True Blue**
VS-6B	**Insignia Red**
VS-9S	**Insignia White**
VS-10S	**Lemon Yellow**
VS-11S	**Willow Green** (Gloss Black in February 1937)
VS-12S	**Gloss Black** (Willow Green in February 1937)
VS-14S	**No Color Used**

After 1 July 1937, the squadrons were redesignated to match the numbering of the Cruiser Scouting Squadrons.

VS-5B	**became**	**VCS-2**
VS-6B	**became**	**VCS-3**
VS-9S	**became**	**VCS-4**
VS-10B	**became**	**VCS-5**
VS-11S	**became**	**VCS-6**
VS-12S	**became**	**VCS-7**
VS-14S	**became**	**VCS-8**

On 1 July 1938, a revision in cruiser tail markings was ordered. Tail markings for the heavy cruiser squadrons retained the single wide stripe across the vertical and horizontal tail surfaces and the light cruiser squadrons were identified with two thinner stripes across the vertical and tail surfaces.The following tail markings were assigned to the VCS squadrons:

VCS-2	**Double True Blue**
VCS-3	**Double Insignia Red**
VCS-4	**Single True Blue**
VCS-5	**Single Lemon Yellow**
VCS-6	**Single Gloss Black**
VCS-7	**Single Willow Green**
VCS-8	**Double Gloss Black**
VCS-9 (1939)	**Double Willow Green**

Three ships not assigned to Cruiser Scouting Squadrons were given the following tail markings:

USS Raleigh CL-7 (Destroyer Flotilla One)	**Double Blue**
USS Detroit CL-8 (Destroyer Flotilla Two)	**Double Blue**
USS Richmond CL-9 (Destroyer Force)	**Double Red**

The aircraft also carried section markings (as listed below) through 26 February 1941. Each section of each squadron had its own ship and color assigned. Section markings were located on the cowl and on a upper wing chevron. The first aircraft in each section also had a fuselage stripe.

A/C No.	BATTLESHIPS and CARRIERS	LIGHT CRUISERS 2-PLANE SECTIONS	HEAVY CRUISERS 4-PLANE SECTIONS
1	Red Stripe & Cowl	Red Stripe & Cowl	Red Stripe & Cowl
2	Red Top Half Cowl	Red Top Half Cowl	Red Top Half Cowl
3	Red Bottom Half Cowl	White Stripe & Cowl	Red Bottom Half Cowl
4	White Stripe & Cowl	White Top Half Cowl	Red Vertical Stripe on Cowl
5	White Top Half Cowl	Blue Stripe & Cowl	White Stripe & Cowl
6	White Bottom Half Cowl	Blue Top Half Cowl	White Top Half Cowl
7	Blue Stripe & Cowl	Black Stripe & Cowl	White Bottom Half Cowl
8	Blue Top Half Cowl	BlackTop Half Cowl	White Vertical Stripe on Cowl
9	Blue Bottom Half Cowl	Green Stripe & Cowl	Blue Stripe & Cowl
10	Black Stripe & Cowl	Green Top Half Cowl	Blue Top Half Cowl
11	BlackTop Half Cowl	Yellow Stripe & Cowl	Blue Bottom Half Cowl
12	Black Bottom Half Cowl	Yellow Top Half Cowl	Blue Vertical Stripe on Cowl
13	Green Stripe & Cowl		Black Stripe & Cowl
14	Green Top Half Cowl		BlackTop Half Cowl
15	Green Bottom Half Cowl		Black Bottom Half Cowl
16	Yellow Stripe & Cowl		Black Vertical Stripe on Cowl
17	Yellow Top Half Cowl		Green Stripe & Cowl
18	Yellow Bottom Half Cowl		Green Top Half Cowl
19			Green Bottom Half Cowl
20			Green Vertical Stripe on Cowl

WWII SOC/SON COLORS

In October 1940, squadrons were directed to replace the aluminum finish with gloss aircraft grey. It is doubtful that any aircraft were repainted before a new 30 December 1940 directive called for aircraft to be painted non-specular light grey overall. Aircraft thus refinished had their fuselage side codes painted in white.

On 13 October 1941, a new directive called for aircraft lower surfaces to be non-specular light grey with all upper surfaces being non-specular blue grey.

On 1 February 1943, the handsome tri-color scheme was ordered. It consisted of non-specular insignia white undersurfaces, non-specular intermediate blue fuselage sides and upper surfaces of semi-gloss sea blue. Propeller tips were orange-yellow with the remainder of the propeller being black.

Although two directives were issued in 1944 to paint carrier aircraft glossy sea blue overall it wasn't until 2 January 1947 that that directive directly applied to the scout/observation aircraft.

SOC/SON TECHNICAL MARKINGS PRE-WAR

A doping code consisting of letters and numbers was employed to designate the material used and the date of application. This doping code information was located on the underside of the fuselage, wings, and control panels. All characters were 1/2 inches high and painted in black. Navy Department specifications assigned the following designating letters to the types of dope authorized:

Acetate Dope	**AD**
Nitrate Dope	**ND**
Semi-Pigmented Nit Dope	**SPN**
Pigmented Acetate Dope	**PA**
Pigmented Nitrate Dope	**PN**

The numbers following the designating letters indicated the number of coats of dope. The concluding series of numbers indicated the date of completion. For example, the code ND2, PN3, 2-8-36 would mean: Nitrate Dope 2-coats, Pigmented Nitrate Dope 3-coats, with the work being completed on 8 February 1936. If a panel needed repair the original data was retained, the date changed and the word "REFINISHED" added. The new doping code was then painted after the original data. For example, ND2, PN3, 7-15-36, REFINISHED SPN2.

All wing struts were numbered to indicate their position on the airplane. The strut number was placed one-inch from the lower end of the strut and a corresponding number was placed on the lower wing or fuselage close to the lower strut fitting. The numbers were one-inch high and painted in black. The front outermost strut on the right wing was numbered one and the remaining front struts were marked from right-to-left with consecutive odd numbers. The right outermost strut was numbered two with the remaining rear struts being marked with consecutive even numbers.

All lift points were indicated by the words "LIFT HERE" in black one-inch letters as was "DISCONNECT ELECTRICAL WIRING BEFORE REMOVING (part)".

SOC-1 LEADING DATA

	FLOATS	WHEELS
LENGTH OVERALL, PLANE LEVEL	31 FT. 8 IN.	26 FT. 10 IN.
HEIGHT OVERALL	14 FT 1 IN.	13 FT. 2 IN.
WING SPAN, UPPER	36 FT.	36 FT.
WING SPAN, LOWER	35 FT. 6 IN.	35 FT. 6 IN.
WING CHORD, UPPER	5 FT. 9 IN.	5 FT. 9 IN.
WING CHORD, LOWER	4 FT. 6 IN.	4 FT. 6 IN.
WING GAP	6 FT. 5 IN.	6 FT. 5 IN.
ANGLE OF INCIDENCE	4°	4°
DIHEDRAL	3.5°	3,5°
STAGGER	20 IN.	20 IN.
SWEEPBACK	4°	4°
WING AREA INCLUDING AILERONS	348 SQ. FT.	348 SQ. FT.
AILERON AREA	31.2 SQ. FT.	31.2 SQ. FT.
STABILIZER AREA	45.5 SQ. FT.	45.5 SQ. FT.
ELEVATOR AREA	25 SQ. FT.	25 SQ. FT.
FIN AREA	10.8 SQ. FT.	10.8 SQ. FT.
RUDDER AREA	14.7 SQ. FT.	14.7 SQ. FT.
AIRFOIL SECTION, UPPER WING	NACA 0010	NACA 0010
AIRFOIL SECTION LOWER WING	NACA 2212	NACA 2212
FUEL TANK CAPACITY	170 GAL.	140 GAL.
OIL TANK CAPACITY	13.5 GAL.	13.5 GAL
NORMAL GROSS WEIGHT	5,153 LBS.	5,000 LBS.
OVERLOAD GROSS WEIGHT	5,341 LBS.	
WEIGHT EMPTY	3,508 LBS.	3.321 LBS.
NORMAL USEFUL WEIGHT	1,645 LBS	1,679 LBS.
CREW WEIGHT	400 LBS.	400 LBS.
ARMAMENT WEIGHT	175 LBS.	175 LBS.
EQUIPMENT WEIGHT	10 LBS.	44 LBS.
RADIO WEIGHT	145 LBS.	145 LBS.
NORMAL LOAD LBS./H.P.	9.6	9.2
NORMAL LOAD LBS./SQ. FT.	15.3	14.6
MAXIMUM SPEED at SEA LEVEL	157.3 M.P.H.	162.2 M.P.H.
MAXIMUM SPEED at 5,000 FEET	153 M.P.H.	158 M.P.H.
MAXIMUM SPEED at 15,000 FEET	133 M.P.H.	139 M.P.H.
STALLING SPEED, NORMAL LOAD	55.9 M.P.H.	55.0 M.P.H.
STALLING SPEED, LESS FUEL	51.1 M.P.H.	50.2 M.P.H.
TIME-TO-CLIMB to 5,000 FEET	5.9 MIN.	5.7 MIN.
TIME-TO-CLIMB to 15,000 FEET	31.1 MIN.	27.0 MIN.
ENDURANCE at 60%v, NORMAL LOAD	8.32 HRS	8.64 HRS
ENDURANCE at 60%v, OVERLOAD	10.0 HRS	
ENDURANCE at 75%v, NORMAL LOAD	5.99 HRS	6.00 HRS
ENDURANCE at 90%v, NORMAL LOAD	3.83 HRS	3.85 HRS
ENDURANCE at MAXIMUM SPEED	2.63 HRS	2.63 HRS
RANGE at ECONOMICAL SPEED, OVERLOAD	954 MILES	
TAKE-OFF RUN, NO WIND		548 FEET
TAKE-OFF RUN, 12 KNOT WIND		297 FEET
TAKE-OFF RUN, 25 KNOT WIND		169 FEET

SOC-1 BuNo 9917 assigned to the USS Raleigh (CL-7), a destroyer division flagship in 1937. (R. Pritchard via Larkins)

CURTISS SOC/SON SQUADRON DESIGNATIONS & AIRCRAFT ASSIGNMENTS

In 1935, the year the SOC first deployed with the fleet, the following battleship observation squadrons and cruiser scout observation squadrons were in service. Squadron designations ending in "B" were for squadrons assigned to battle forces and those ending in "S" were assigned to scouting forces.

COMMANDER BATTLESHIPS, BATTLE FORCE

Battleship Division One	VO-1B
Battleship Division Two	VO-2B
Battleship Division Three	VO-3B
Battleship Division Four	VO-4B
COMMANDER CRUISERS, BATTLE FORCE	
Cruiser Division Two	VS-5B
Cruiser Division Three	VS-6B
COMMANDER CRUISERS, SCOUTING FORCE	
Cruiser Division Four	VS-9S
Cruiser Division Five	VS-10S
Cruiser Division Six	VS-11S
Cruiser Division Seven	VS-12S

SOC AIRCRAFT ASSIGNMENTS JUNE 1936

COMMANDER CRUISERS, BATTLE FORCE		
Cruiser Division Two	VS-5B	6 SOC-1
Cruiser Division Three	VS-6B	8 SOC-1
COMMANDER CRUISERS, SCOUTING FORCE		
Cruiser Division Four	VS-9S	17 SOC-1
Cruiser Division Five	VS-10S	12 SOC-1
Cruiser Division Six	VS-11S	13 SOC-1
Cruiser Division Seven	VS-12S	10 SOC-1
SCOUTING FORCE		
USS Indianapolis CA-35		3 SOC-1
COMMANDER, SPECIAL SERVICE SQUADRON		
USS Memphis CL-13		2 SOC-1
COMMANDER DESTROYERS, SCOUTING FORCE		
USS Raleigh CL-7		2 SOC-1
COMMANDER AIRCRAFT, BASE FORCE		
Flag Unit		2 SOC-1
ASIATIC FLEET		
USS Augusta CA-31		4 SOC-1
Navy Yard, Cavite, P.I.		2 SOC-1
COMMANDER AIRCRAFT, BATTLE FORCE		
USS Saratoga CV-3 Utility Unit		1 SOC-1

SOC AIRCRAFT ASSIGNMENTS JUNE 1937

COMMANDER BATTLESHIPS, BATTLE FORCE		
Battleship Division Two	VO-2B	1 SOC-1
Battleship Division Four	VO-4B	1 SOC-1
COMMANDER CRUISERS, BATTLE FORCE		
Cruiser Division Two	VS-5B	6 SOC-1
Cruiser Division Three	VS-6B	6 SOC-1, 1 SOC-2
COMMANDER DESTROYERS, BATTLE FORCE		
USS Detroit CL-8		1 SOC-1, 1 SOC-2
BATTLE FORCE POOL		
NAS San Diego		2 SOC-1
NAS Norfolk		1 SOC-1
COMMANDER CRUISERS, SCOUTING FORCE		
ComCruScoFor Flag Unit		1 SOC-1
Cruiser Division Four	VS-9S	11 SOC-1, 6 SOC-2
Cruiser Division Five	VS-10S	13 SOC-1, 3 SOC-2
Cruiser Division Six	VS-11S	12 SOC-1, 1 SOC-2
Cruiser Division Seven	VS-12S	8 SOC-1, 4 SOC-2
SCOUTING FORCE		
USS Indianapolis CA-35		4 SOC-1
SCOUTING FORCE POOL		
NAS Norfolk		18 SOC-1, 8 SOC-2
		(4 for Boise, 4 for Brooklyn)
SPECIAL SERVICE SQUADRON		
USS Omaha CL-4		2 SOC-1
USS Erie PG-50		1 SOC-2
USS Charleston PG-51		1 SOC-1
COMMANDER DESTROYERS, SCOUTING FORCE		
USS Raleigh CL-7		4 SOC-1
COMMANDER AIRCRAFT, BASE FORCE		
Flag Unit		2 SOC-1
USS Langley AV-3		3 SOC-2
USS Wright AV-1		1 SOC-2
ASIATIC FLEET		
USS Augusta CA-31		3 SOC-1
Navy Yard, Cavite, P.I.		3 SOC-1
FLEET AIR BASE, COCO SOLO		
Utility Unit		1 SOC-1
COMMANDER AIRCRAFT, BATTLE FORCE		
Flag Unit		2 SOC-1
NAVAL AIRCRAFT FACTORY STORAGE		
For USS Philadelphia CL-41		4 SOC-2
NAS ANACOSTIA, D.C.		
VX-4D5		2 SOC-1, 1 SOC-2
NAS NORFOLK		
Repair and storage		1 XSOC-1, 2 SOC-2

SQUADRON REORGANIZATION 1 JULY 1937

The suffix letter "B" was deleted and Battleship Observation Squadrons became VO-1 through VO-4. The Cruiser Scouting Squadrons were re-numbered to conform with the number of the cruiser divisions and the suffix letters "B" and "S" were deleted. The letter "C" was added to the squadron designation for Cruiser Scouting Squadrons and the "S" on the fuselage side became "CS". Thus the following changes were dictated:

VS-5B	became	VCS-2
VS-6B	became	VCS-3
VS-9S	became	VCS-4
VS-10B	became	VCS-5
VS-11S	became	VCS-6
VS-12S	became	VCS-7
VS-14S	became	VCS-8

SOC AIRCRAFT ASSIGNMENTS JUNE 1938

UNITED STATES FLEET	
CinCUS Flag Ship	1 SOC-3
USS Pennsylvania BB-38 ship unit	3 SOC-3
BATTLE FORCE	
ComBatFor Flag Unit	1 SOC-3
USS California BB-44 ship unit	3 SOC-3

COMMANDER BATTLESHIPS, BATTLE FORCE

Battleship Division One	VO-1	6 SOC-3
Battleship Division Two	VO-2	6 SOC-3
Battleship Division Three	VO-3	9 SOC-3
Battleship Division Four	VO-4	6 SOC-1, 3 SOC-3

COMMANDER CRUISERS, BATTLE FORCE

Cruiser Division Two	VCS-2	6 SOC-1
Cruiser Division Three	VCS-3	8 SOC-1
Cruiser Division Eight	VCS-8	2 SOC-2, 8 SOC-3
Cruiser Division Nine	VCS-9	4 SOC-3
USS Medusa (repair ship)	AR-1	2 SOC-1

COMMANDER DESTROYERS, BATTLE FORCE

CO Flotilla One USS Raleigh CL-7		2 SOC-1
CO Flotilla Two USS Detroit CL-8		2 SOC-1

COMMANDER CRUISERS, SCOUTING FORCE

ComCruScoFor Flag Unit		1 SOC-1
Cruiser Division Four	VCS-4	12 SOC-1, 2 SOC-2
Cruiser Division Five	VCS-5	16 SOC-1
Cruiser Division Six	VCS-6	10 SOC-1, 1 SOC-2
Cruiser Division Seven	VCS-7	13 SOC-1, 3 SOC-2

COMMANDER AIRCRAFT, SCOUTING FORCE

Flag Unit		1 SOC-1, 1 SOC-2

PATROL WING ONE

USS Langley AV-3		3 SOC-2
USS Wright AV-1		1 SOC-2

ASIATIC FLEET

USS Augusta CA-31		5 SOC-1
Navy Yard, Cavite, P.I.		3 SOC-1

FLEET AIR BASE, SAN PEDRO

Commander Base Force		1 SOC-2

SUBMARINE FORCE

USS Richmond CL-9		2 SOC-1

COMMANDER AIRCRAFT, BATTLE FORCE

Flag Unit Carrier Division One		1 SOC-1, 1 SOC-3
Flag Unit Carrier Division Two		1 SOC-1

SOC AIRCRAFT ASSIGNMENTS JUNE 1939

UNITED STATES FLEET

CinCUS Flag Ship		1 SOC-3
USS Pennsylvania BB-38 ship unit		3 SOC-3

BATTLE FORCE

ComBatFor Flag Unit		1 SOC-3
USS California BB-44 ship unit		3 SOC-3

COMMANDER BATTLESHIPS, BATTLE FORCE

Battleship Division One	VO-1	6 SOC-3
Battleship Division Two	VO-2	6 SOC-3
Battleship Division Three	VO-3	6 SOC-3, 3 SON-1
Battleship Division Four	VO-4	7 SOC-3, 2 SON-1

COMMANDER CRUISERS, BATTLE FORCE

USS Saint Louis CL-49		4 SON-1
Cruiser Division Two	VCS-2	1 SOC-1, 2 SOC-2
Cruiser Division Eight	VCS-8	7 SOC-1, 9 SOC-3
Cruiser Division Nine	VCS-9	9 SOC-1, 6 SOC-2 1 SOC-3, 4 SON-1

COMMANDER DESTROYERS, BATTLE FORCE

CO Flotilla One USS Raleigh CL-7		2 SOC-1
CO Flotilla Two USS Detroit CL-8		2 SOC-1

SCOUTING FORCE

ComCruScoFor Flag Unit		1 SOC-3
USS Indianapolis CA-35		2 SOC-1, 2 SOC-2

COMMANDER CRUISERS, SCOUTING FORCE

Flag Unit		1 SOC-1
USS Wichita (CA-45)		4 SOC-1
Cruiser Division Four	VCS-4	12 SOC-1, 4 SOC-2
Cruiser Division Five	VCS-5	11 SOC-1, 1 SOC-2
Cruiser Division Six	VCS-6	11 SOC-1, 2 SOC-2
Cruiser Division Seven	VCS-7	13 SOC-1, 3 SOC-2

COMMANDER AIRCRAFT, SCOUTING FORCE

Flag Unit		2 SOC-3
PATROL WING ONE		
ComPatWingOne		1 SOC-2
USS Langley AV-3		1 SOC-1, 1 SOC-2, 1 SON-1
USS Wright AV-1		1 SOC-1
PATROL WING TWO		1 SOC-2
PATROL WING FOUR		1 SOC-1
PATROL WING FIVE		
ComPatWingFive		1 SON-1

ASIATIC FLEET

USS Augusta CA-31		4 SOC-1
USS Marbelhead CL-12		2 SOC-1

FLEET AIR BASE, SAN PEDRO

Commander Base Force		1 SOC-1

SUBMARINE FORCE

USS Richmond CL-9		2 SOC-2

COMMANDER AIRCRAFT, BATTLE FORCE

Flag Unit Carrier Division One		1 SOC-1, 2 SOC-3
Flag Unit Carrier Division Two		1 SOC-1
USS Wasp CV-7		3 SON-1

SOC AIRCRAFT ASSIGNMENTS JUNE 1940

UNITED STATES FLEET

CinCUS Flag Ship		1 SOC-3
USS Pennsylvania BB-38 ship unit		3 SOC-3

BATTLE FORCE

ComBatFor Flag Unit		1 SOC-3
USS California BB-44 ship unit		3 SOC-3

COMMANDER BATTLESHIPS, BATTLE FORCE

Battleship Division One	VO-1	6 SOC-3
Battleship Division Two	VO-2	5 SOC-3
Battleship Division Three	VO-3	6 SOC-3, 3 SON-1
Battleship Division Four	VO-4	7 SOC-3, 2 SON-1
Battleship Division Five	VO-5	4 SOC-3, 5 SON-1

COMMANDER CRUISERS, BATTLE FORCE

Cruiser Division Three	VCS-3	3 SOC-1, 2 SOC-2 1 SOC-3, 1 SON-1
Cruiser Division Eight	VCS-8	6 SOC-1, 5 SOC-3 6 SON-1
Cruiser Division Nine	VCS-9	1 SOC-1, 5 SOC-3 11 SON-1

COMMANDER DESTROYERS, BATTLE FORCE

CO Flotilla One USS Raleigh CL-7		2 SOC-3
CO Flotilla Two USS Detroit CL-8		2 SOC-1

COMMANDER CRUISERS, SCOUTING FORCE

ComCruScoFor Flag Unit		1 SOC-3
USS Indianapolis CA-35		3 SOC-1, 1 SOC-2
Cruiser Division Four	VCS-4	13 SOC-1, 3 SOC-2
Cruiser Division Five	VCS-5	14 SOC-1, 2 SOC-2
Cruiser Division Six	VCS-6	13 SOC-1, 5 SOC-2

Cruiser Division Seven	VCS-7	9 SOC-1, 7 SOC-2

COMMANDER AIRCRAFT, SCOUTING FORCE

USS Memphis CL-13	1 SOC-1, 1 SOC-3

PATROL WING ONE

ComPatWingOne	1 SOC-1

PATROL WING TWO

ComPatWingTwo	1 SOC-1
Pearl Harbor	1 SOC-2
Utility Unit	1 SOC-1

PATROL WING FIVE

USS Patoka AO-9	2 SOC-1, 1 SON-1

ASIATIC FLEET

USS Augusta CA-31	3 SOC-1, 1 SOC-2
USS Marblehead CL-12	2 SOC-1
USS Langley AV-3	2 SOC-1, 1 SOC-2, 1 SON-1

SUBMARINE FORCE

USS Richmond CL-9	2 SOC-2

SPECIAL SERVICE SQUADRON

USS Charleston PG-51	1 SOC-1
USS Erie PG-50	1 SOC-1

COMMANDER AIRCRAFT, BATTLE FORCE

Flag Unlt	2 SOC-3
Flag Unit Carrier Division One	1 SOC-1
USS Saratoga CV-3	2 SOC-3
USS Lexington CV-2	3 SOC-1
Carrier Division Two	
USS Yorktown CV-5	2 SOC-2
USS Enterprise CV-6	1 SOC-3

ATLANTIC FLEET

USS Ranger CV-4	2 SOC-1
USS Wasp CV-7	1 SOC-1, 2 SON-1
Aviation Unit USS Noa DD-343	1 XSOC-1, 1 SON-1

SOC ASSIGNMENTS 31 DECEMBER 1941

SOC-1

NAS Alameda	5	NAS Anacostia	2
NAS Pensacola	8	Norfolk, Lant	8
San Diego Scout Force	3	NAS San Diego	1
Pearl Harbor reserve	7	NAS Seattle	1
USS Astoria CA-34	4	NAS Bermuda	1
USS Erie PG-50	1	Alameda for Bat. Force	4
USS Houston CA-30	1	USS Saratoga CV-3	4
NAS Quonset Point	1	USS Augusta CA-31	3
USS Charleston PG-51	1	USS Chester CA-27	2
USS Chicago CA-29	1	USS Louisville CA-28	3
USS Ranger CV-4	2	USS Noa DD-343	1
USS Yorktown CV-5	1	USS Minneapolis CA-36	4
USS New Orleans CA-32	2	USS Northampton CA-26	4
USS Pensacola CA-24	4	USS Portland CA-33	3
USS Indianapolis CA-35	4	USS Salt Lake City CA-25	4
USS San Francisco	2	USS Omaha CL-4	1
USS Quincy CA-39	3	USS Langley AV-3	1
USS Cincinnati CL-6	1	USS Marbelhead CL-12	2
Patrol Wing Two	1	Navy Yard Cavite	2
Total 103 SOC-1s			

SOC-1A

VS-201 USS Long Island AVG-1	1

SOC-2

USS Louisville CA-28	1	USS Saratoga CV-3	1
ComScoFor	1	NAS Anacostia	2
NAS Pensacola	1	Norfolk, Lant	4
Pearl Harbor Bat Force	4	ConAirLant	1
NAS Alameda	2	USS Chester CA-27	2
USS New Orleans	2	USS Portland CA-33	1
USS Quincy CA-39	1	USS Wichita CA-45	2
USS Augusta CA-31	1	USS CHICAGO CA-29	3
USS San Francisco	3	USS Nashville CL-43	1
USS Langley AV-3	1	USS Vincennes CA-44	1
Navy Yard Cavite	1	Total 36 SOC-2s	

SOC-3

Alameda Bat Force	5	NAS Norfolk	1
Norfolk, Lant	4	USS Milwaukee CL-5	1
USS Boise CL-47	3	USS Concord CL-10	1
USS Enterprise CV-6	2	USS Detroit CL-8	1
USS Honolulu CL-48	3	USS Helena CL-50	3
USS Omaha CL-4	1	NAS San Juan, PR	1
USS Philadelphia CA-41	3	NAS San Diego	1
USS Phoenix CL-46	4	USS Raleigh CL-7	2
USS Savannah CL-42	1	USS Saint Louis CL-49	3
Pearl Harbor Bat Force	4	USS Trenton CL-11	2
USS Tuscaloosa CA-37	1	ComAirLant	1
USS Brooklyn CL-40	1	USS Wichita CA-45	1
USS Vincennes CA-44	3	USS Lexington CV-2	1
USS Houston CL-30	2	Navy Yard Cavite	2
Total 59 SOC-3s			

SOC-3A

Norfolk. Lant	1	VS-201 USS Long Island	12
ACTU, Norfolk	1	Total 14 SOC-3As	

SON-1

USS Saratoga CV-3	3	USS Honolulu CL-48	1
NAS Anacostia	2	Norfolk, Lant	4
NAS Alameda	2	USS Brooklyn CL-40	3
USS Boise CL-47	1	USS Cincinnati CL-6	1
USS Detroit CL-8	1	USS Philadelphia CL-41	1
USS Houston CA-30	1	USS Nashville CL-43	3
Pearl Harbor Bat Force	1	USS Helena CA-75	1
USS Savannah CL-42	3	USS Richmond CL-9	2
USS Tuscaloosa CA-37	3	USS Wasp CV-7	1
USS Wichita CA-45	1	USS Memphis CL-13	1
USS Saint Louis CL-49	1	USS Concord CL-10	1
Total 38 SON-1s		Total 251 SOC/SONs	

FEBRUARY 1946

The only SOC/SONs on hand in February 1946 were seven assigned to SOSU-3 at NAS Alameda, CA. These were: 2 SOC-1s, 3 SOC-3s and 2 SON-1s.

Below, mystery plane with one-off unusual fuselage markings of number 163. Date, location and unit are unknown. (SDAM)

CURTISS SOC-1/-2/-3/SON-1 PILOT'S MAIN INSTRUMENT PANEL

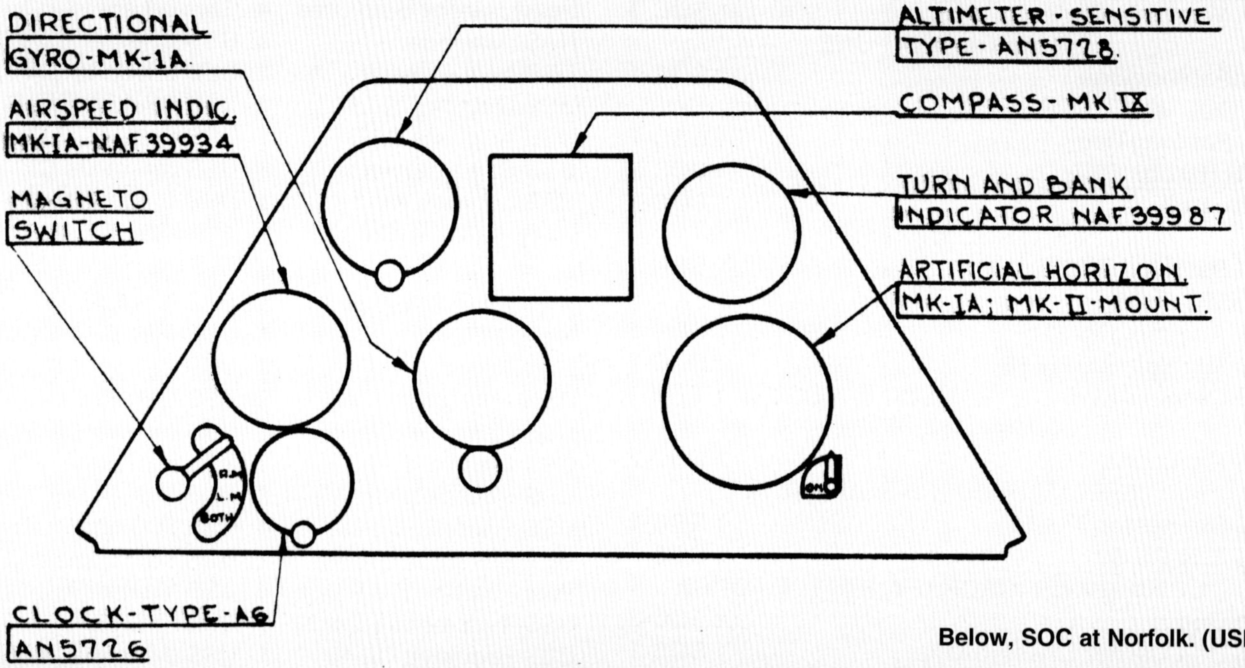

Below, SOC at Norfolk. (USN)

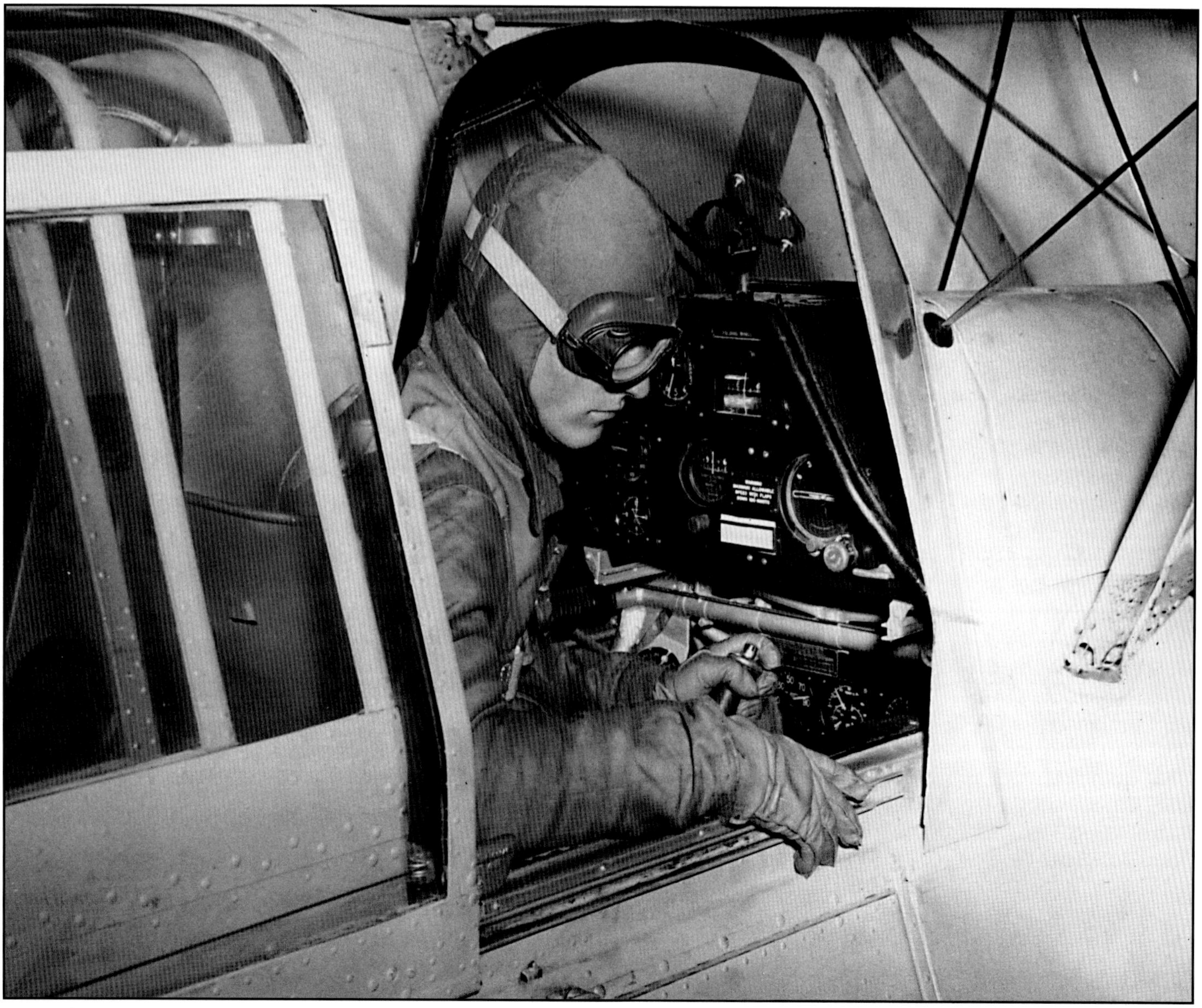

CURTISS SOC-1/-2 PILOT'S AUXILIARY INSTRUMENT PANEL

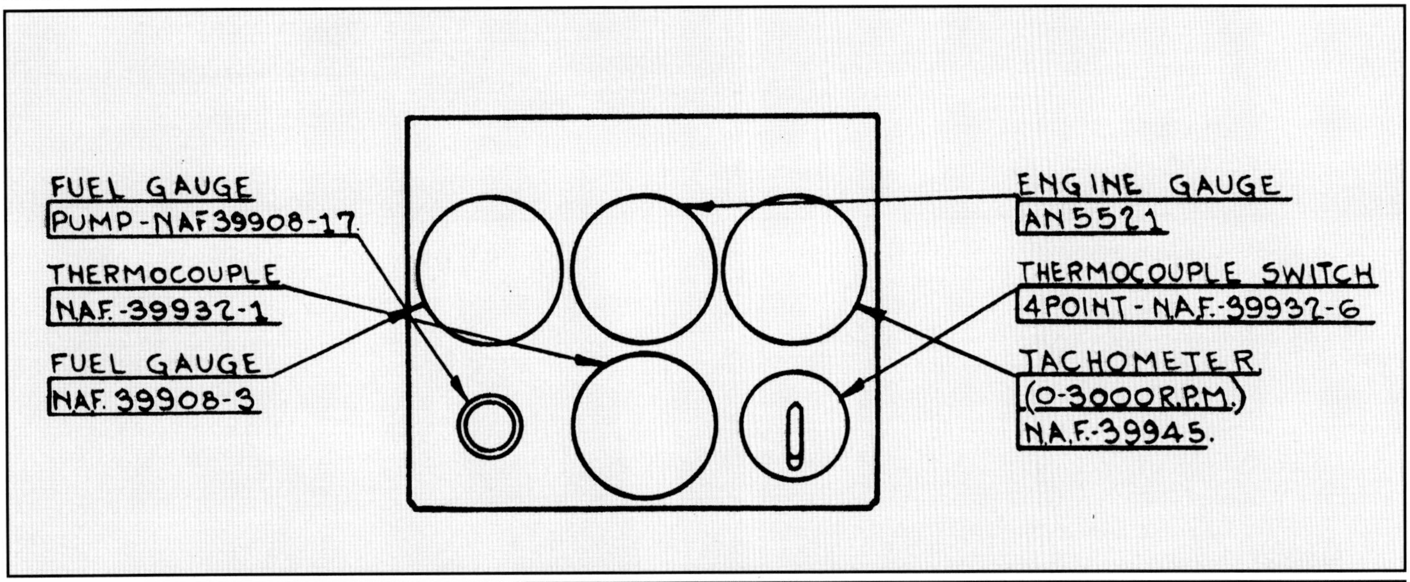

CURTISS SOC-3/-4 PILOT'S AUXILIARY INSTRUMENT PANEL

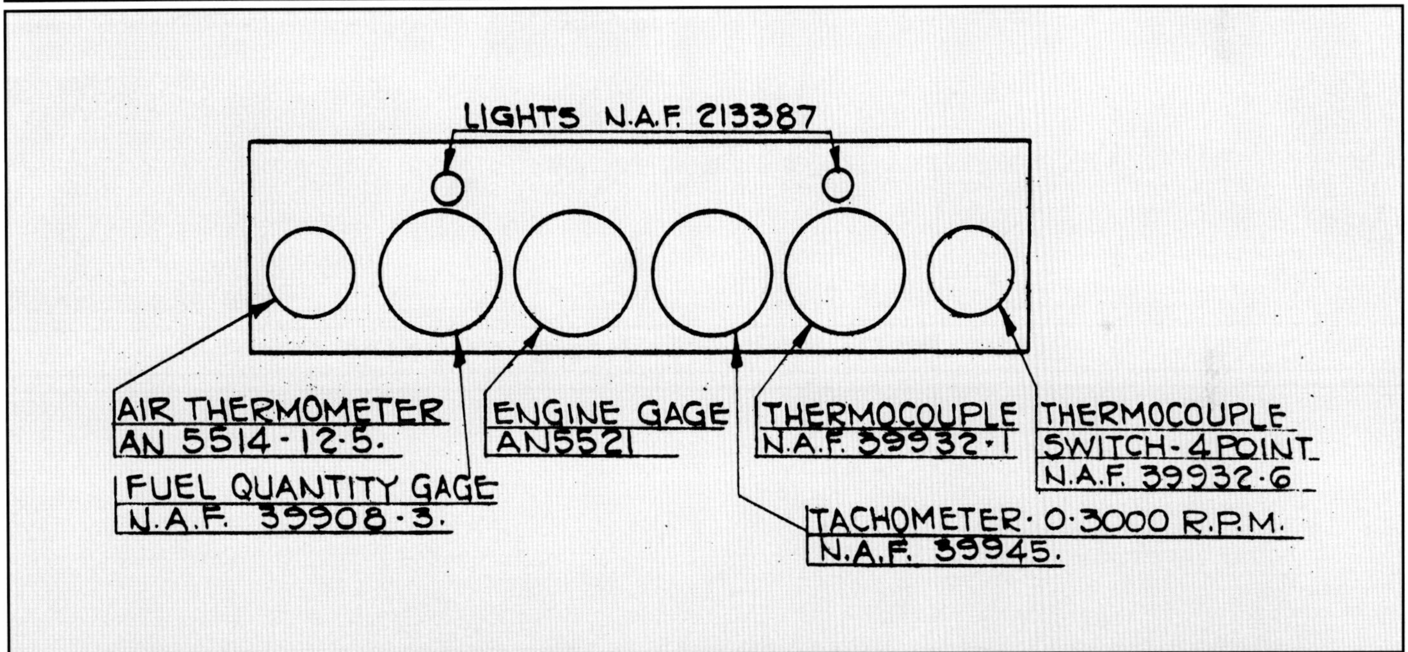

CURTISS SOC-1/-2/-3/-4/SON-1 OBSERVER'S INSTRUMENT PANEL

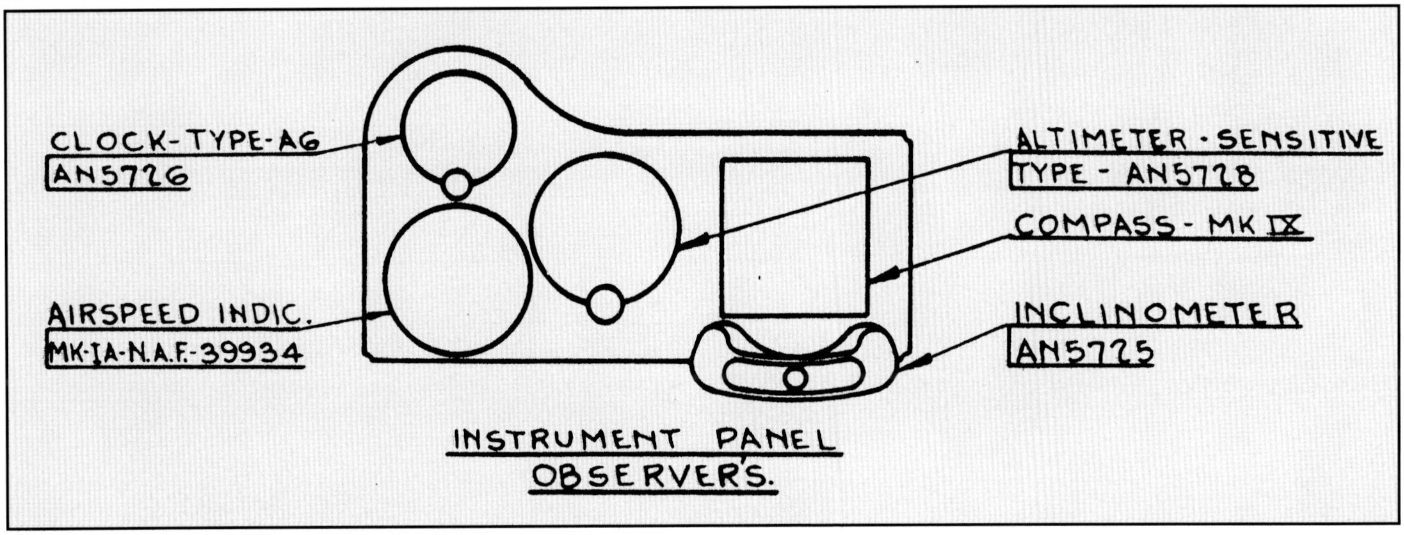

SOC-1 PILOT'S COCKPIT

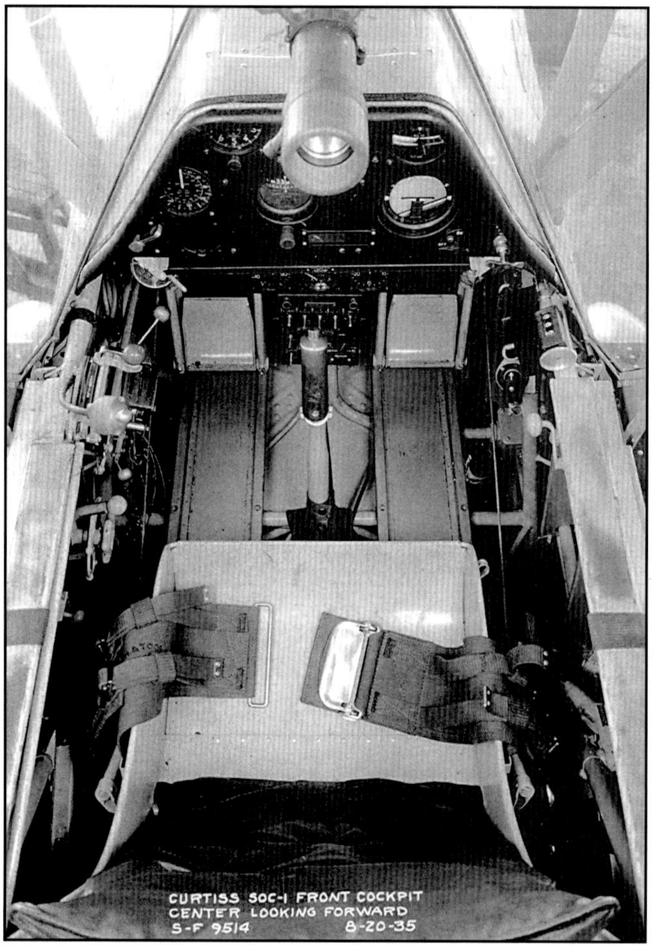

SOC-1/-2/-3/-4/SON-1 PILOT'S SEAT

SOC-1 PILOT'S RIGHT - HAND COCKPIT

CURTISS SOC-2 PILOT'S INSTRUMENT PANEL

CURTISS SOC-2 PILOT'S LEFT - HAND COCKPIT

CURTISS SOC-2 PILOT'S RIGHT - HAND COCKPIT

CURTISS SOC-3 PILOT'S COCKPIT

CURTISS SOC-3 PILOT'S LEFT - HAND COCKPIT

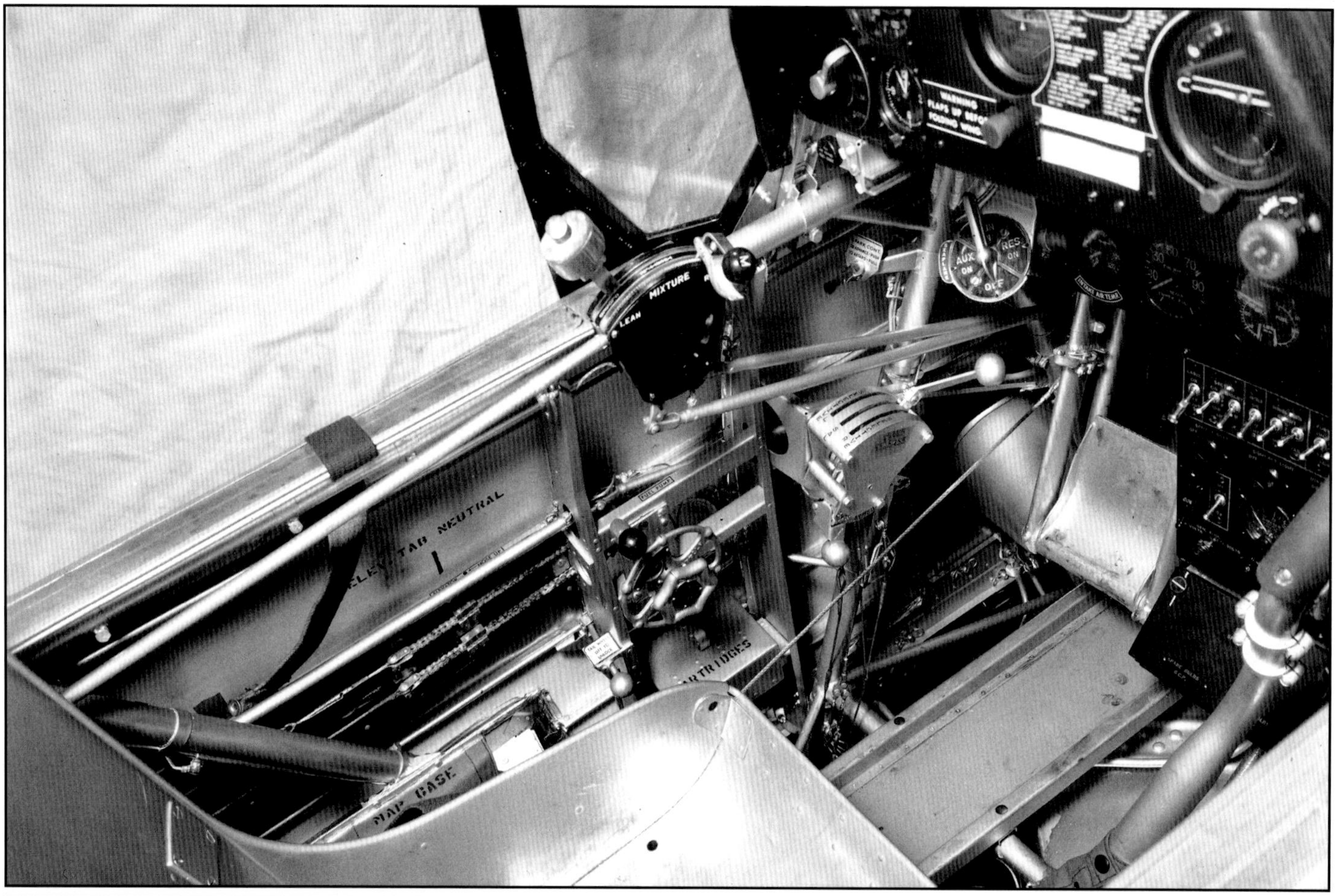

CURTISS SOC-3 PILOT'S RIGHT - HAND COCKPIT

USCG CURTISS SOC-4 PILOT'S COCKPIT

CURTISS SOC/SON AFT COCKPIT

Above, typical SOC-1/3 aft right-hand cockpit on 24 December 1937. (USN) Typical SOC-1/3 aft left-hand cockpit on 24 December 1937. (USN)

Above, SOC-1 aft cockpit looking forward on 16 August 1935. (USN) Above right, SOC-1 aft cockpit with mock up radio gear installed on 20 August 1935. (USN) Below, SOC-1 aft cockpit from above with .30 cal. machine gun mounted on 16 August 1935.

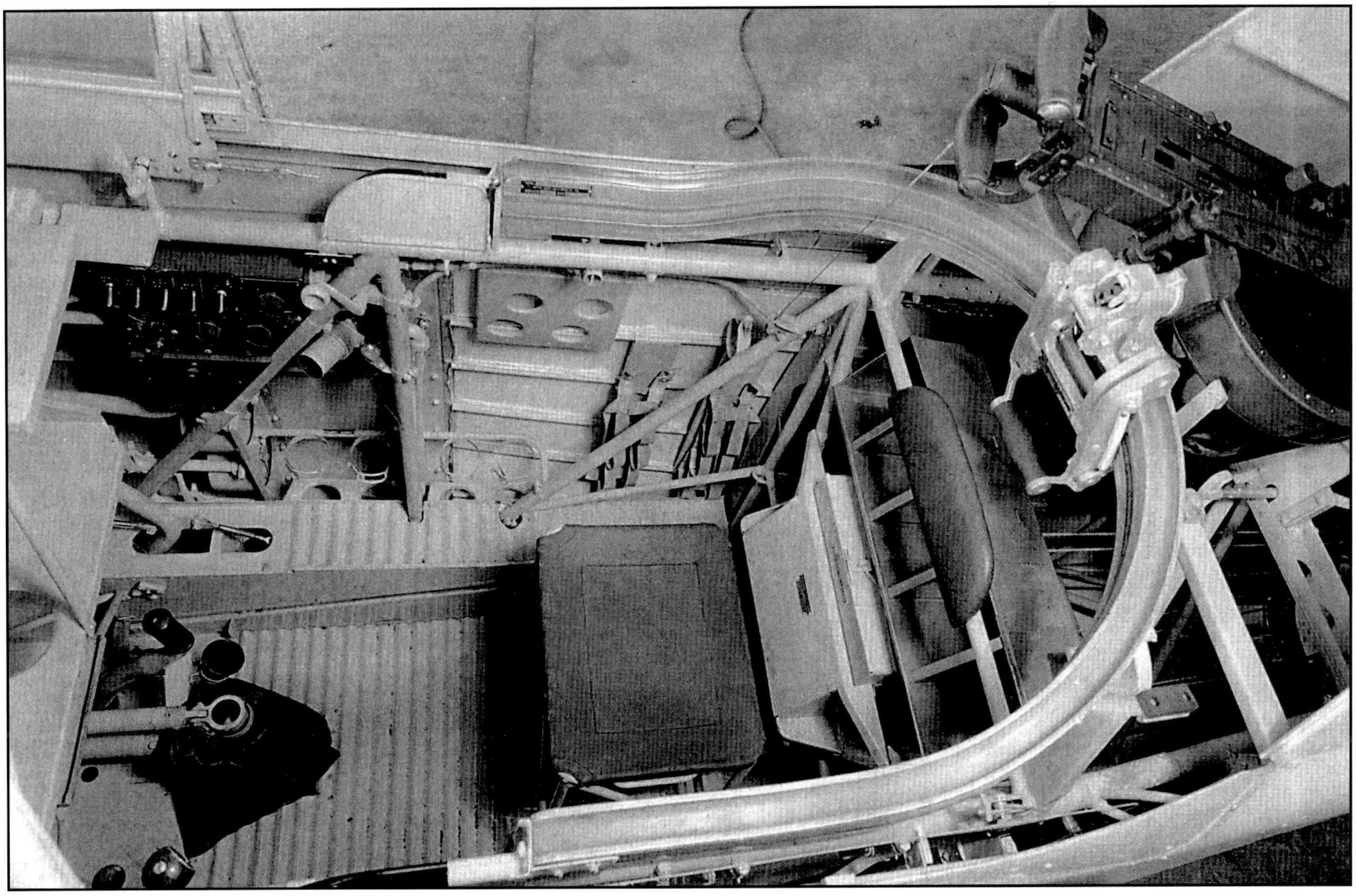

USCG CURTISS SOC-4 AFT COCKPIT, NO GUN POSITION

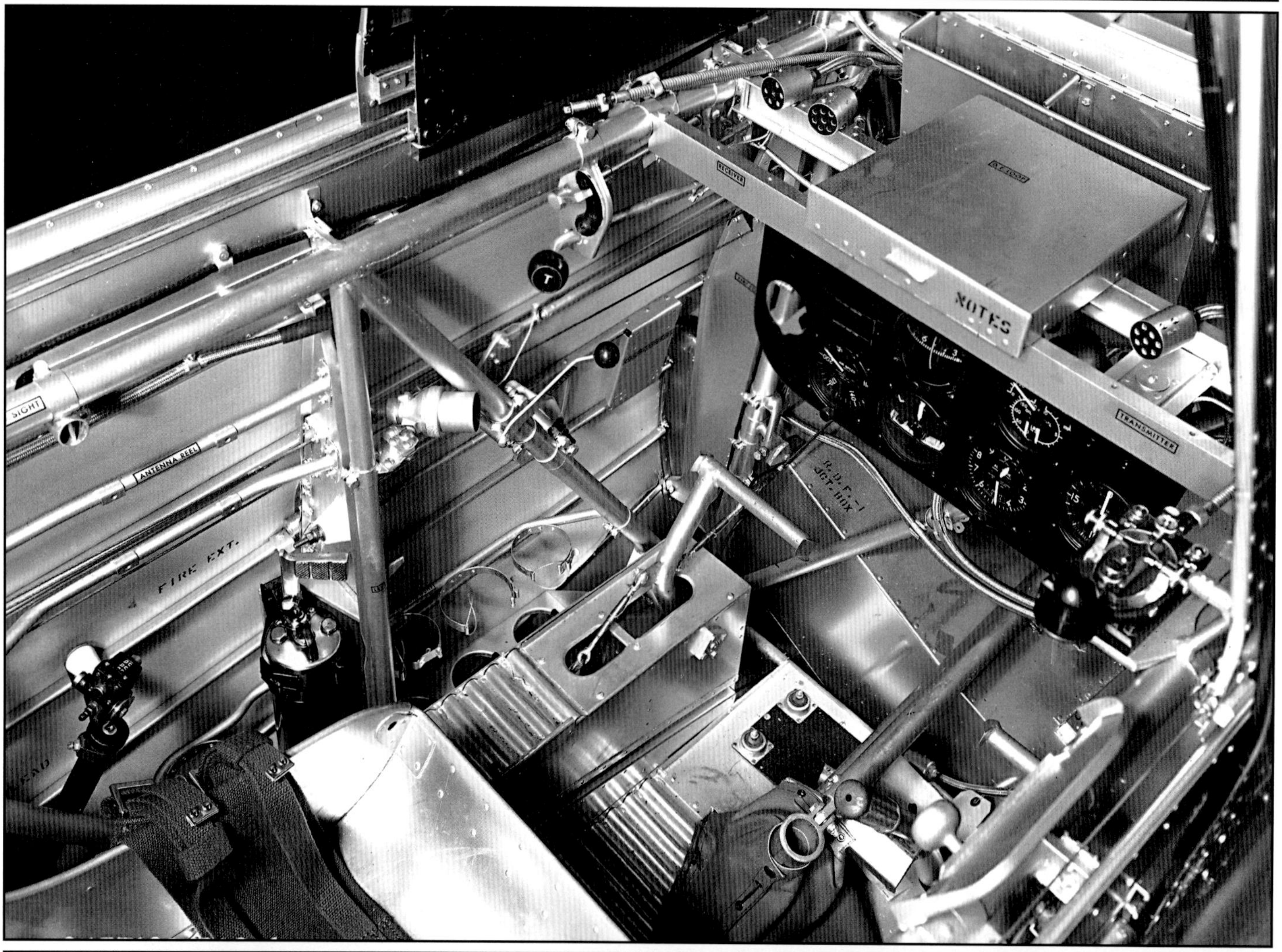

DEPTH BOMBS

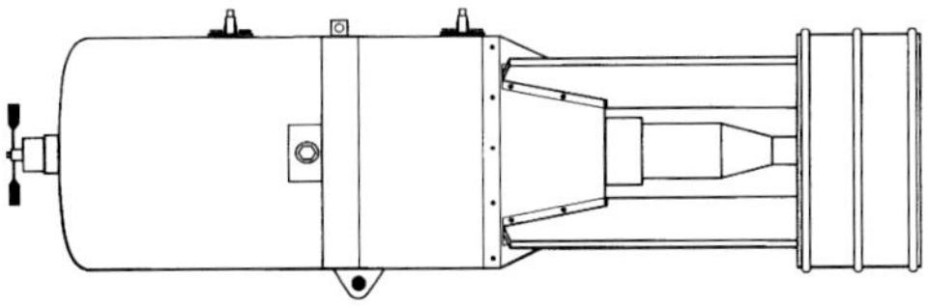

Mk.53 325-pound depth bomb

WWII depth bombs were filled with 70% of either TNT, torpex, or HBX. They were produced in four sizes: 325, 350, 650 and 700 pounds. The SOC could only field the 325 and 350lb versions.

Loading or unloading a Mk.17 depth bombs on an SOC aboard the USS Philadelphia (CL-41) in 1942. (USN)

CURTISS SOC/SON .30 CALIBER FLEXIBLE MACHINE GUN

At left, turtledeck lowered and machine gun deployed. (USN) Above, turtledeck lowered and machine gun stowed externally. (USN) Below, machine gun stowage inside the raised turtledeck on an SOC-1 on 16 August 1935. (USN) The SOC-4 was not built with an aft gun position.

CURTISS SOC/SON .30 CALIBER FIXED BROWNING M2 MACHINE GUN

Fixed gun seen from above.

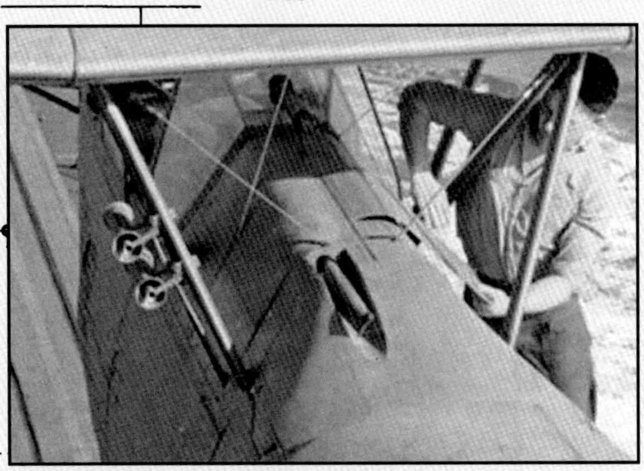

Pilot's fixed forward firing Browning M2 machine gun with 500 rounds was fired through the propeller arc. It was complemented by a gun camera and a Mark III Mod. 2 telescope sight.

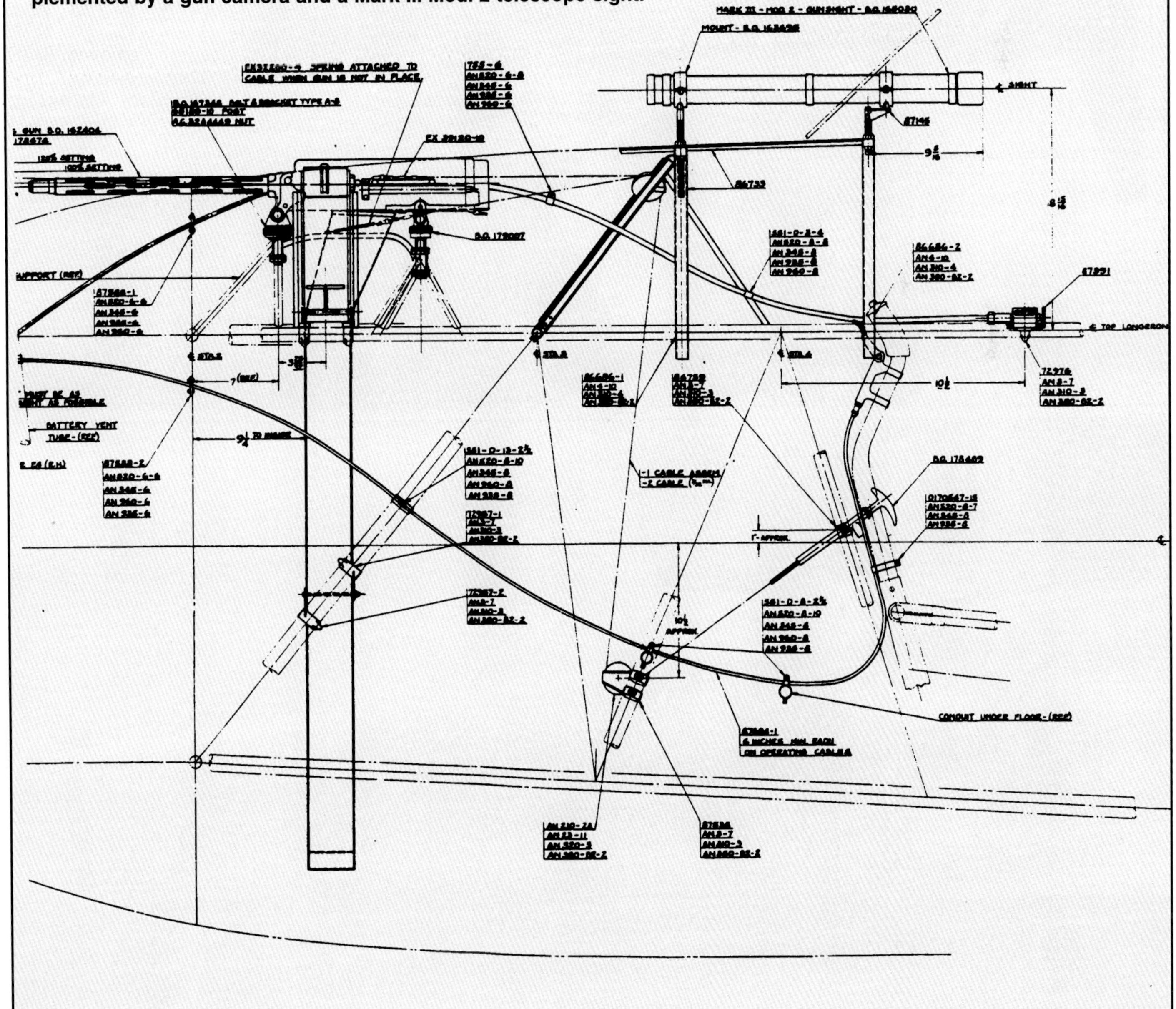

CURTISS SOC-3 CONTRACT 54679
FUSELAGE ASSEMBLY-L.H. FRONT HALF
S-F 10821 3-3-38

CURTISS SOC-3 CONTRACT 54679
FUSELAGE ASSEMBLY-R.H. FRONT HALF
S-F 10823 3-3-38

CURTISS WRIGHT R-1340 ENGINE INSTALLATION

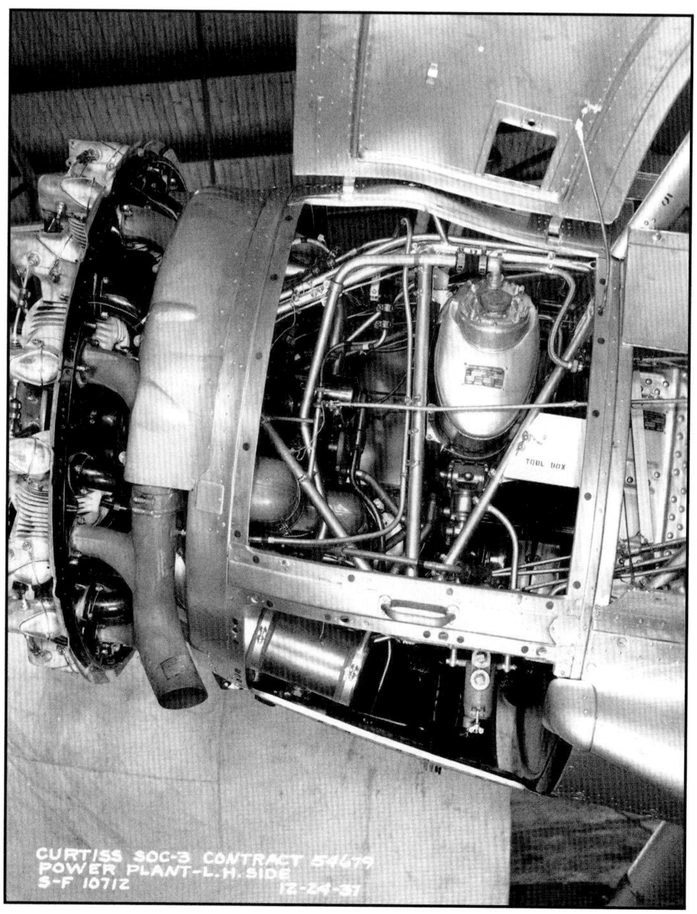

Above, first SOC-1 BuNo 9856 being weighed. (USN) Below, SOC-1 BuNo 9885 during final assembly. (USN)

CURTISS SOC/SON VERTICAL TAIL AND RUDDER

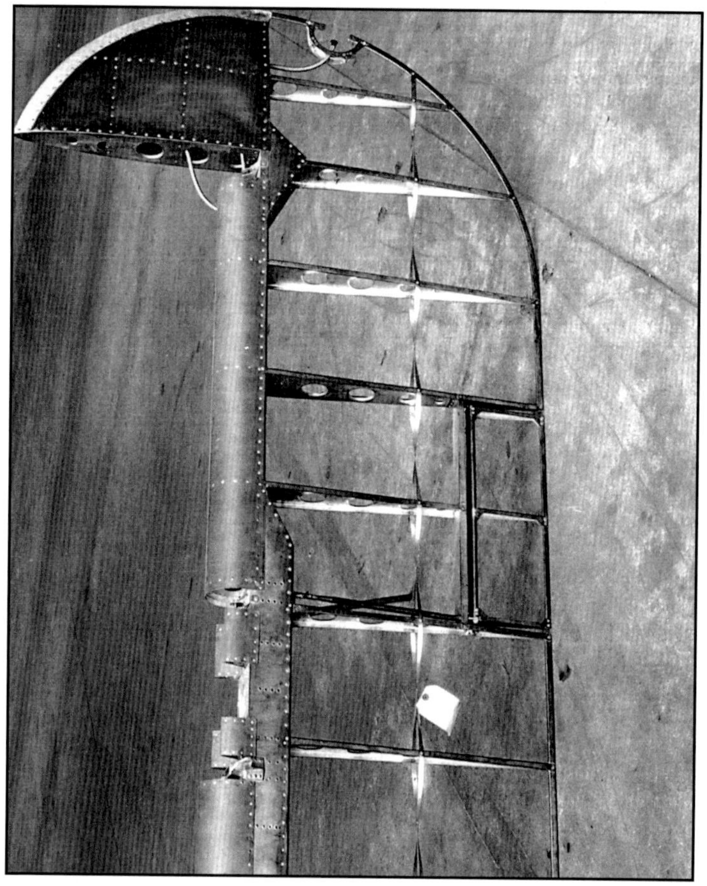

Above, completed SOC-1 rudder (typical all versions). Above right, rudder skeleton prior to application of fabric. Below left, metal vertical fin with skin removed from one side. Below right, vertical tail assembly dimensions. (USN)

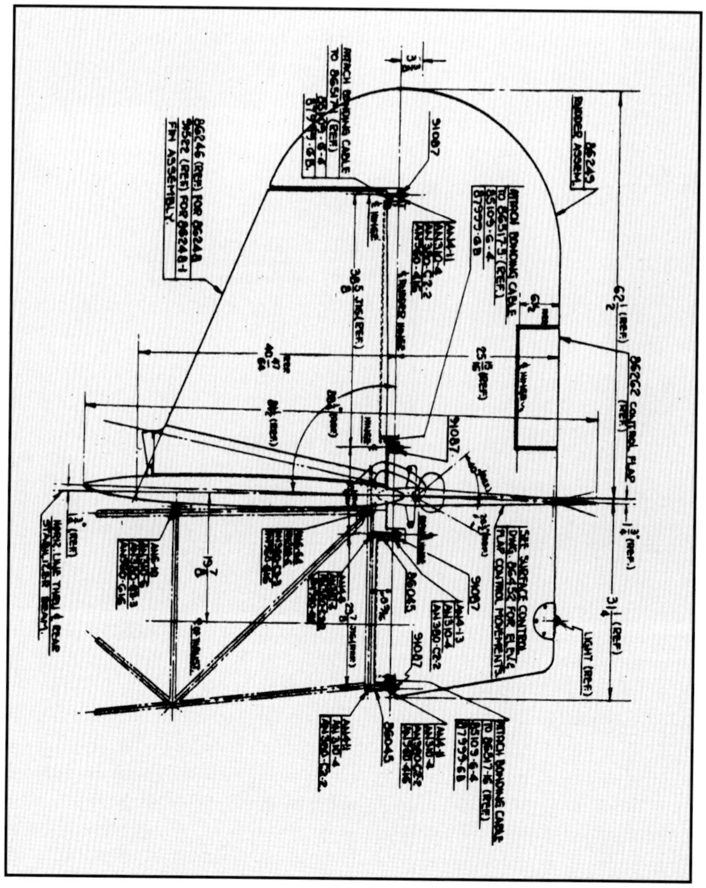

CURTISS SOC/SON HORIZONTAL TAIL AND ELEVATOR

HORIZONTAL TAIL BRACE

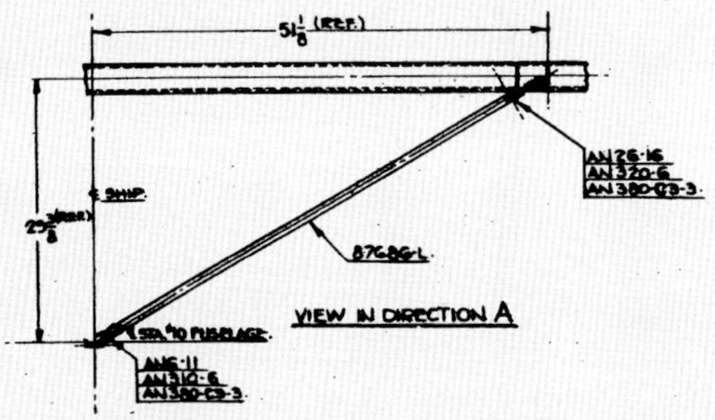

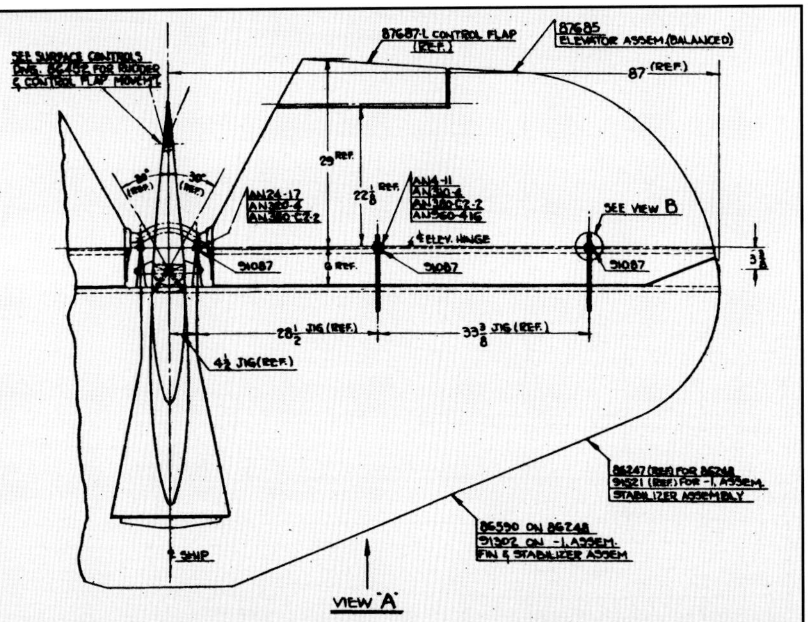

Below, horizontal stabilizer with the topside skin removed. Bottom, elevator assembly skeleton before canvas was applied. (USN)

WING DIMENSIONS, RIGGING

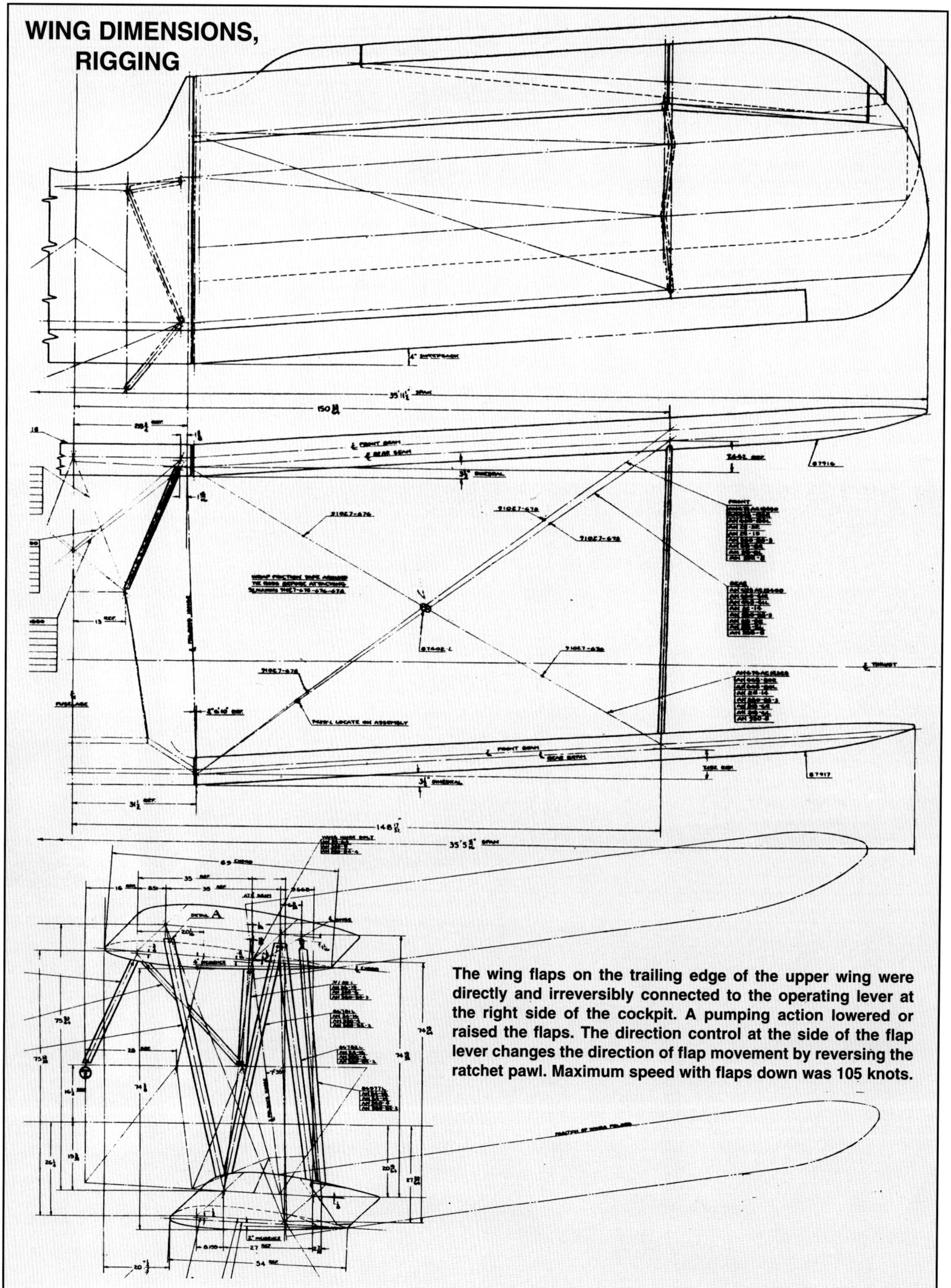

The wing flaps on the trailing edge of the upper wing were directly and irreversibly connected to the operating lever at the right side of the cockpit. A pumping action lowered or raised the flaps. The direction control at the side of the flap lever changes the direction of flap movement by reversing the ratchet pawl. Maximum speed with flaps down was 105 knots.

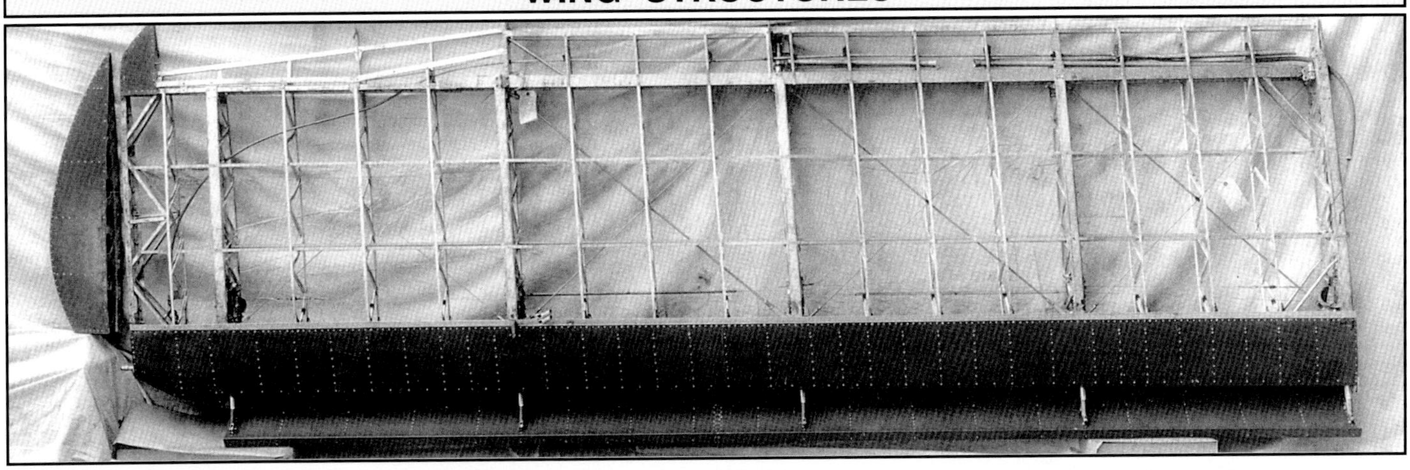

At top, upper wing panel skeleton, slots open seen from below. Above, upper wing panel skeleton, slots open seen from above. At right, upper wing panel skeleton assembly, slots closed. Bottom, finished fabric-covered upper wing panel. (USN)

WING STRUCTURES

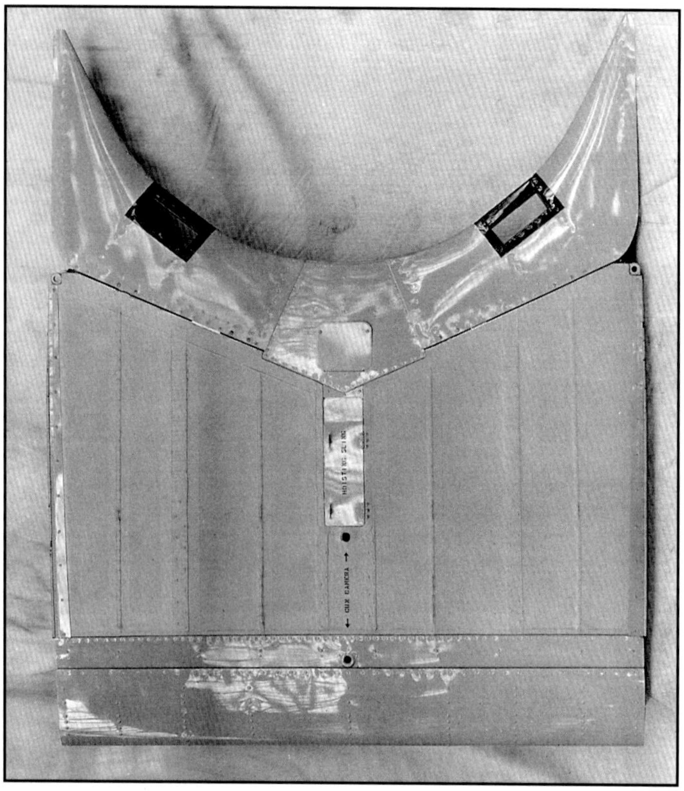

Above, upper wing center section finished with hoisting sling panel closed. Above right, upper wing center section without fabric skin and showing hoisting sling. Below, lower wing skeleton. Bottom, upper side of a finished lower wing. (USN)

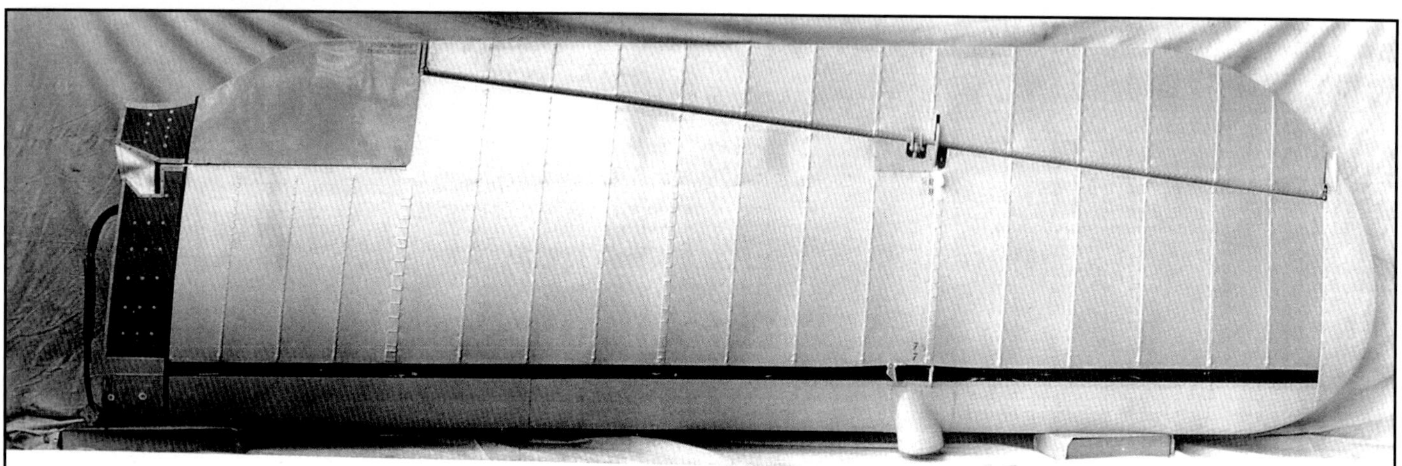

CURTISS SOC/SON MAIN FLOAT

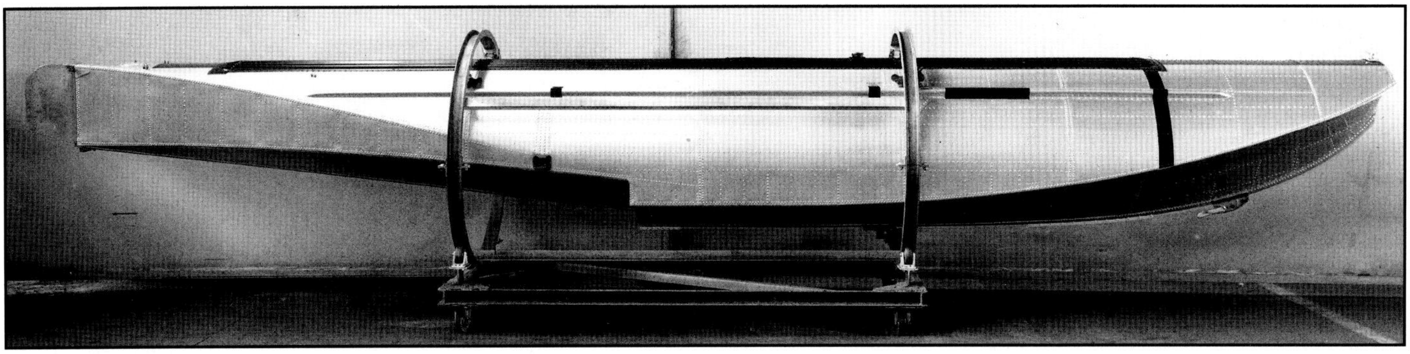

Above, finished main float. (USN) Below, main float fuel tank installation. (USN) Below right, main float water

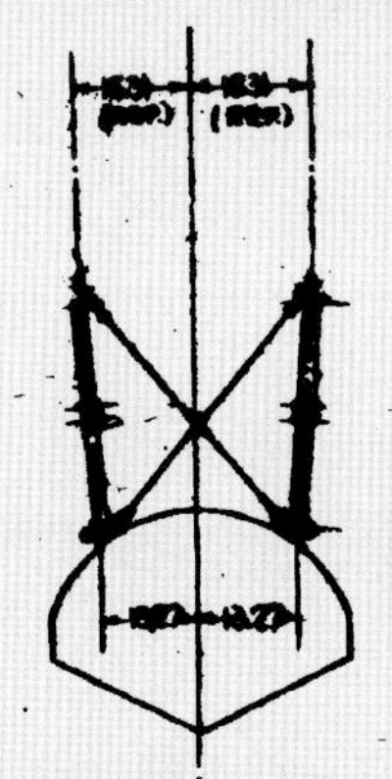

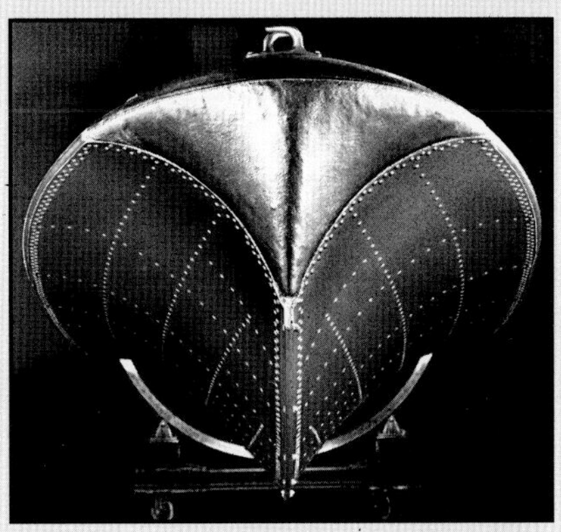

The water rudder was controlled directly from the rudder pedals and could be connected or disconnected by the engaging control lever located at the left side of the pilot's seat. The water rudder was disconnected when the lever was vertical and engaged when the lever was horizontal. It was disconnected before making a water landing.

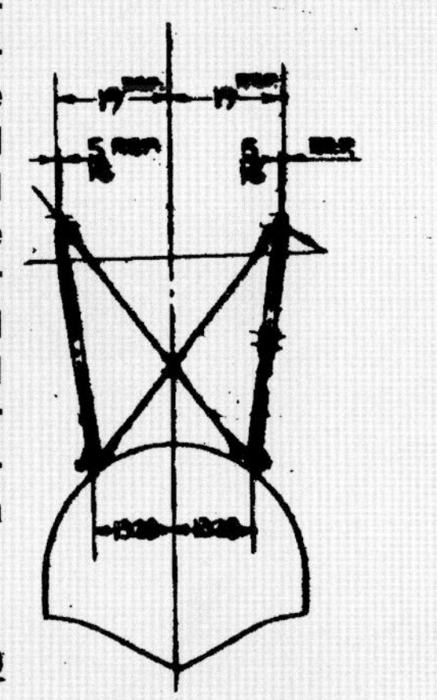

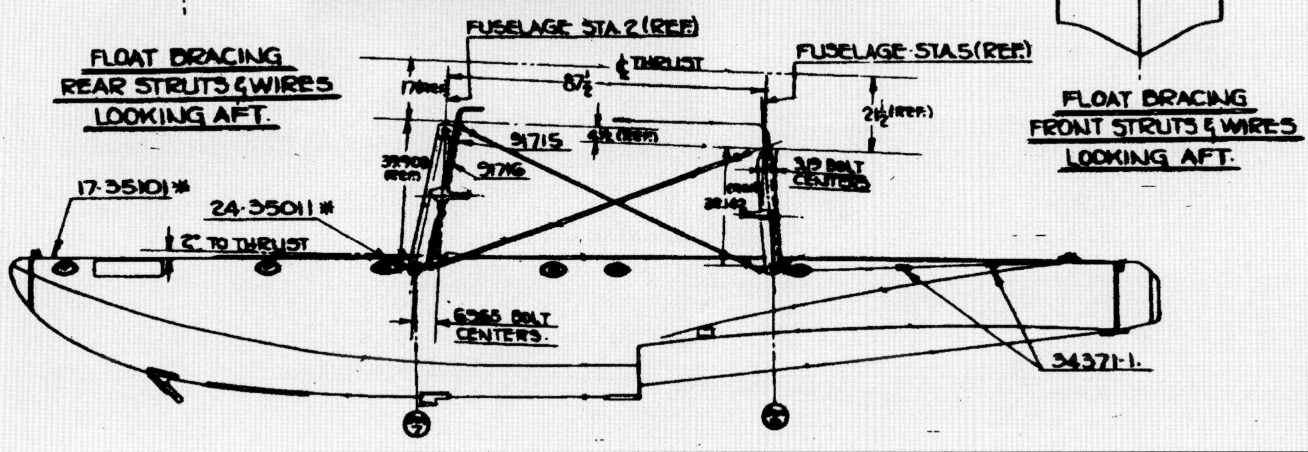

CURTISS SOC/SON WING - TIP FLOATS

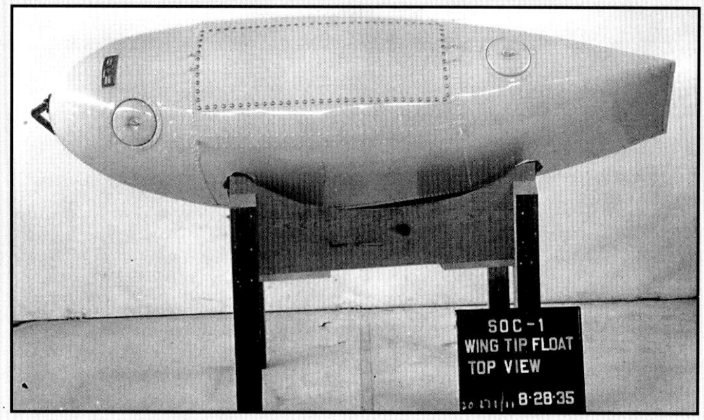

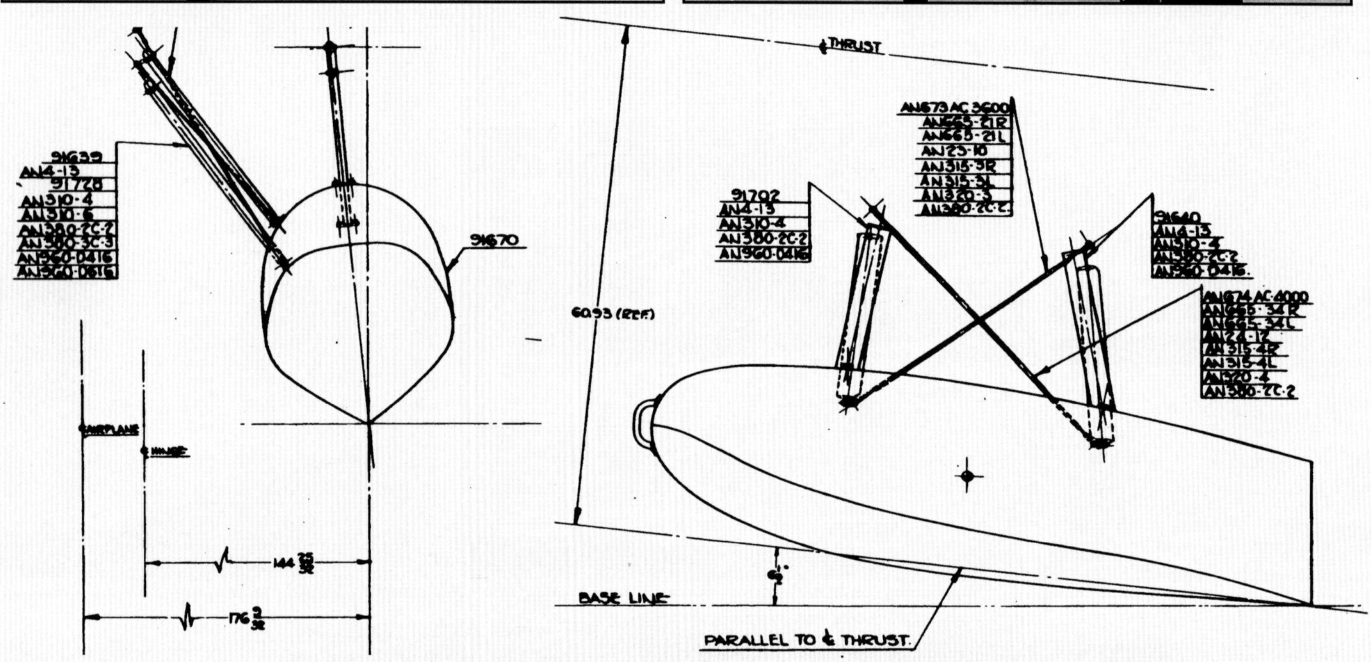

Below, SOC-3 BuNo 1064 on beaching dolly in 1938 shows float layout and rigging. (USN)

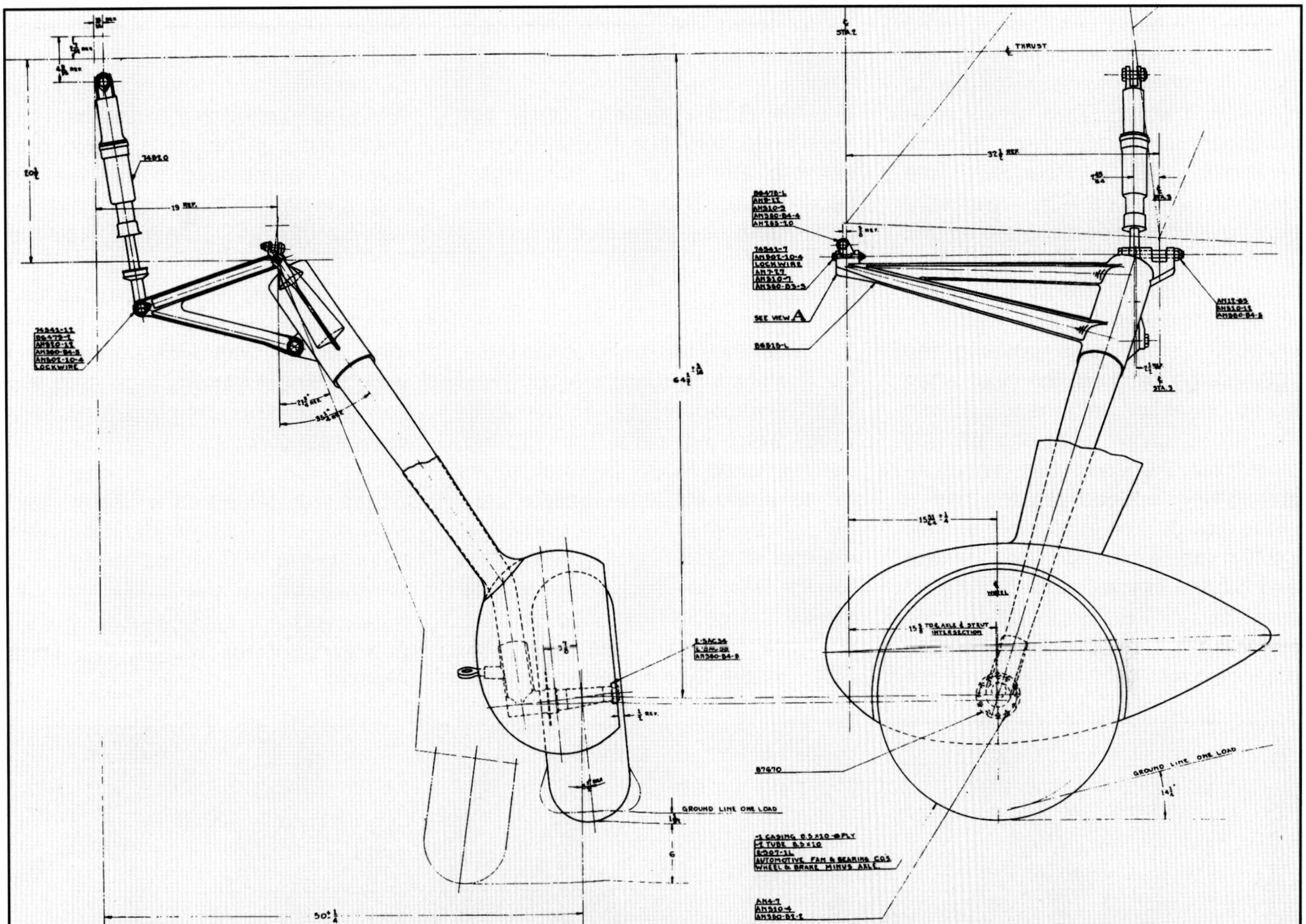

The tail wheel lock was located on the left side of the cockpit. When the handle was raised to the upper notch, the tail wheel was free to swivel 360°. With the handle in the down position the tail wheel would automatically lock when it trailed in the line of flight. It had a centering device to make it trail aft when off the ground.

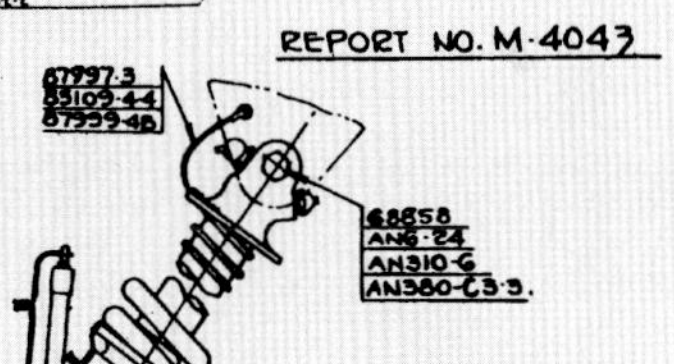

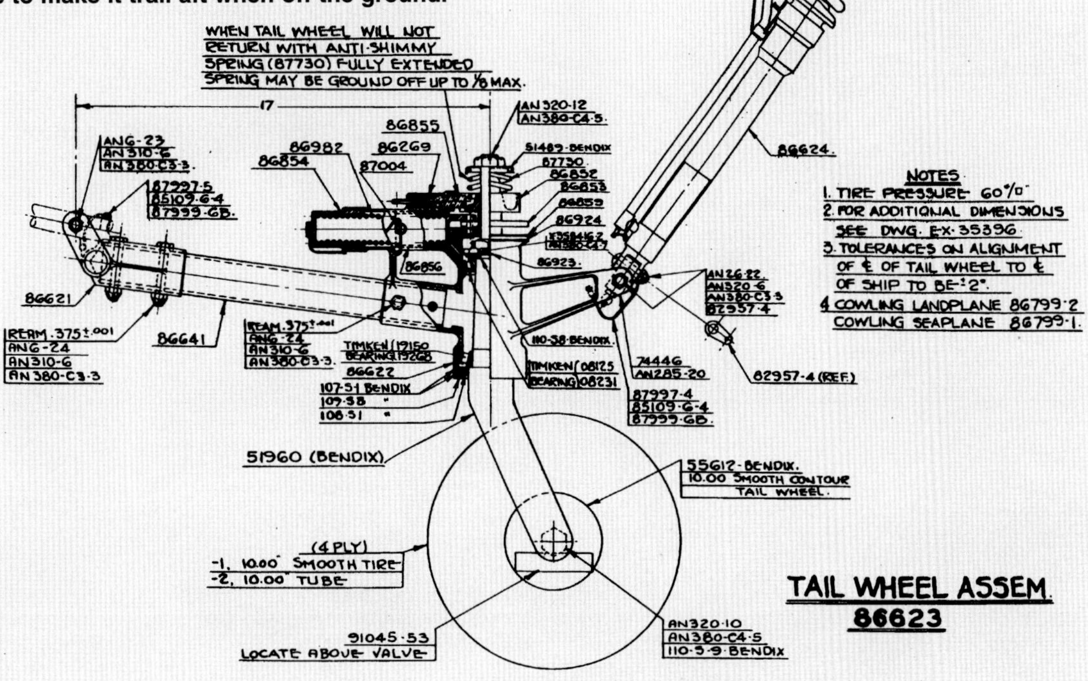

TWO NAVY PILOTS MAKE MID-AIR RESCUE of MARINE PARACHUTIST

On 15 May 1941, an SOC-3 became the rescue vehicle for the most incredible mid-air rescue in NAVAIR history. Two newspaper headlines on the 16th were "Miracle in Midair" and "A Rescue of the Doomed". A Marine R2D-1 was conducting a practice parachute drop and 11 men and one cargo pack had jumped when a cargo pack was dropped followed by 2Lt. Walter S. Osipoff. Osipoff was pushing the 150lb cargo pack out the door when his rip chord line became fouled with that of the cargo pack and he was pulled out the door. His parachute immediately opened but the pack's chute did not. The strain on the static cable lines caused the forward attachment bracket and bracing to carry away. The cargo chute then opened and Osipoff's chute became entangled with the static line and the 13 rip chord lines leaving him dangling head first 100ft aft of the aircraft.

The R2D-1 slowed down and returned to North Island while six passengers attempted to pull Osipoff back inside. This effort was abandoned and they attempted to lower a parachute which was also abandoned.

While the R2D-1 circled the base, LT W.W. Lowrey (base test pilot) and Chief Aviation Pilot J.R. McCants were watching from the ground. Lowrey asked McCants "Shall we try it?" to which McCants replied "Let's try". They manned an unmarked refurbished SOC-3 and taxied off the line. A Marine ran alongside and tossed a large knife into the aft cockpit. After joining with the R2D-1, the SOC made 3-or-4 rescue passes before McCants was able to grab Osipoff and hold him atop the canopy between the front and rear cockpits. McCants slashed at the rip chords but could not cut the metal static line. Lowrey then decided to cut the static line with the SOC's prop. While he was inching the SOC higher to cut the lines, the two planes hit some rough air and the lines were severed as was 12-inches of the R2D-1's tail. Lowrey then landed with McCant standing and holding Osipoff's legs atop the canopy while his torso and head were in the rear cockpit.

Lowrey retired as a Captain and McCants as a Commander. Both men received the DFC for the rescue.

Below, NAP J.R. McCants and LT W.W. Lowrey after the rescue. (via Mark Aldrich)

MARINE SOCs

The Marines had only one squadron of SOCs, it was Marine Scouting Squadron Two (VMS-2) stationed at North Island, CA.

At right, VMS-2 SOC-3 near San Diego on 31 May 1940. (USMC) Below, VMS-2 SOC-3s BuNos 1127 (2-MS-7) and 1128 at Union Air Terminal, Burbank, CA, in June 1939. Aircraft have floatation gear attached to the forward fuselage. (William L. Swisher) Bottom, SOC-3 2-MS-1. (via Mark Aldrich)

NEUTRALITY PATROL MARKINGS

On 19 March 1940, a directive was issued for the markings of Neutrality Patrol aircraft. The National Insignia was relocated to both sides of the forward fuselage and had a diameter of 24-inches on the SOCs. The aircrafts' rudders were painted in three vertical stripes painted fore-to-aft blue, white, red. Larger patrol aircraft used a 36-inch National Insignia on the forward fuselage.

Above, SOC with neutrality patrol star on forward fuselage but no rudder stripes. (USN) At right, VCS-4 SOC-2s on USS Augusta with neutrality patrol rudder stripes and painted-out forward fuselage National Insignias. (USN) Below, SOC-1 BuNo 9877 assigned to Patrol Wing Five in neutrality patrol markings. Seen here with forward mounted National Insignias and blue, white, red rudder stripes at Logan Field, Baltimore, MD. The aircraft was assigned to the seaplane tender USS Patoka AV-6/AO-9 in early 1940. (Berb Ederr via Dave Lucabaugh)

Above, SOC-2 BuNo 0421 assigned to the USS Trenton (CL-11). (via Peter M. Bowers) VCS-4 SOC-2s assigned to the USS Augusta (CA-31) at anchor in Shanghai in August 1937 with neutrality patrol markings applied to the rudders and with the forward fuselage National Insignia painted out. (USN)

EARLY SOC CRASHES

Above, SOC crash at Elizabeth City, NC, in 1941. (via Mark Aldrich) Below, salvage boat recovers a damaged SOC assigned to the USS Augusta (CA-31) while assigned to the Asiatic Fleet. (via Mark Aldrich)

SOC COMMAND AIRCRAFT / FLAGPLANE

Above, SOC-1 BuNo 9877 staff aircraft for Commander Cruisers Scouting Force. Fuselage, landing gear and wing struts were Admiral blue, upper surface of top wing chrome yellow, and the remaining surfaces aluminum. (Gordon S. Williams) Below, BuNo 9877 at Union Air Terminal, Burbank, CA, in 1937. (William Swisher) SOC Flagplanes were assigned to: CIC U.S. Fleet, CDR Battle Force, CDR Cruisers Scouting Force, CDR Carrier Division One, CDR Carrier Division Two, CDR Carrier Division Three, CDR Patrol Patrol Wing Two, CDR Patrol Patrol Wing Five, CDR Patrol Patrol Wing Ten, CDR Scouting Force, CDR Aircraft Asiatic Fleet, CDR Aircraft Atlantic Fleet, CDR Aircraft Base Force, CDR Aircraft Battle Force, and CDR Aircraft Scouting Force.

Above, SOC-3 BuNo 1079 Flagplane at Oakland in June 1938. It belonged to ADM Claude C. Bloch, Commander-In-Chief of the U.S. Fleet. It was the only Command Plane with a blue tail. U.S. Fleet was painted in white below the aft cockpit and U.S. Navy was painted below the horizontal stabilizer. Wings and floats were silver, fuselage Admiral blue and top of upper wing orange yellow. (William T. Larkins) Below, Admiral blue and silver SOC-1 BuNo 9870 assigned to NAS San Pedro in 1938. (USN)

REAR ADMIRAL

VICE ADMIRAL

Above, RADM E.J. King debarks from an SOC-1 Flagplane on the USS Lexington in June 1936. (USN) Bottom, all-blue SOC-3 BuNo 1079 Flagplane of the Commander-In-Chief of the U.S. Fleet aboard his Flagship the USS Pennsylvania (BB-38). (USN)

Above, patch for the aviation unit aboard the USS Noa (DD-343). (via Tailhook)

Above, SOC-2 assigned to the USS Langley (AV-3) in 1938. (via Mark Aldrich)
Below, SOC-2 BuNo 0395 assigned to the USS Erie (PG-50). (via Mark Aldrich)

BATTLESHIP OBSERVATION SQUADRON TWO, VO-1B/VO-1

Above, VO-1B SOC-3 BuNo 1084 assigned to the USS Arizona (BB-39) on 15 February 1938 while at the factory. Tail and upper half of cowl were red. (USN) At right, VO-1 SOC-3 BuNo 1076 assigned to the USS Pennsylvania (BB-38). Below, SOC-3 BuNo 1081 assigned to the USS Nevada (BB-36) at NAS Oakland, CA, in 1938. The upper half of the engine cowl was white. (William T. Larkins) VO-1B/VO-1 SOC tails were red.

VO-1B/VO-1

At left, VO-1 insignia had red lightning bolts and a red chevron. The gun turret was black and the ship's name was painted on the turret. The banner was red. (NAH via Tailhook).

Below, VO-1 SOC-3 BuNo 1077 assigned to the USS Pennsylvania. (BB-38) in 1938 at NAS Oakland, CA. (William T. Larkins) Bottom, VO-1 SOC-3 BuNo 1076 from the Pennsylvania on the seaplane ramp at Oakland in 1938. The upper half of the engine cowl was blue. Note the end of the retractable fairlead antenna on the fuselage side below the number one. (William T. Larkins)

BATTLESHIP OBSERVATION SQUADRON TWO, VO-2B/VO-2

VO-2/VO-2B's insignia on an aircraft. The large background disc was red and the water blue. (via Tailhook)

Above, VO-2B SOC-3s at San Diego, CA, in 1838. (USN) Below, VO-2 SOC-3 during water taxi. Upper half of engine cowl was blue. (USN)

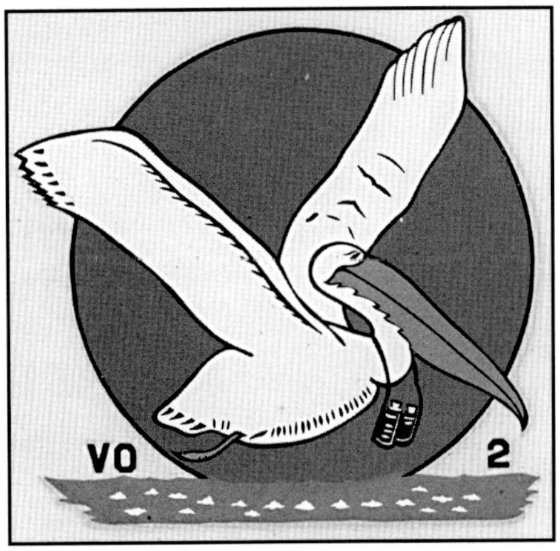

Below, VO-2 SOC-3 BuNo 1065 assigned to the USS Oklahoma (BB-37) on 23 December 1937. Cowl, fuselage stripe and tail were white. (USN)

BATTLESHIP OBSERVATION SQUADRON THREE, VO-3

VO-3/3B insignia pictures "Oswald, the Lucky Rabbit" with red shirt and blue pants riding a large caliber shell. (Tailhook Association) Above, VO-3B SOC-3 BuNo 1086 with "Oswald the Rabbit" insignia on the nose was assigned to the USS New Mexico (BB-40) on 15 February 1938. Cowl, fuselage stripe and tail were blue. (USN) At right, VO-3 SON-1 during night maintenance. (via Mark Aldrich) Below, VO-3 SOC-3 assigned to the USS Mississippi (BB-41) was seen at Bremerton, WA, in 1938. (USN)

BATTLESHIP OBSERVATION SQUADRON FOUR, VO-4

Above, VO-4 SOC-3 BuNo 1110 assigned to the USS West Virginia (BB-48) and flown by CO LCDR Norman Hitchcock. Engine cowl and fuselage stripe was red, tail was black. (USN) Below, VO-4 SOC-1 BuNo 9939 is the backdrop for a unit photo of the USS West Virginia's Marine Detachment. (USN)

Above, VO-4 SOC ready to launch from afterdeck catapult on the USS West Virginia (BB-48). Lower half of cowl was red and the tail was black. (USN) Below, West Virgina-based 4-O-3 launching off the fantail. (USN)

Above, VO-4 insignia. The battleship, smoke and gunfire were black, the "O" was red, the "V" was yellow and the water was blue. (Tailhook)

At right and below, VO-4 SOC aircraft aboard the USS West Virginia (BB-48). (USN)

VO-4

At right, good view of upper wing chevron on a VO-4 SOC. The aircraft number 13 indicates the section leader from section five. Section five chevron color was green bordered by black. (USN) Below, VO-4 SOC-3 BuNo 1115 assigned to the USS Maryland (BB-46) at NAS Oakland, CA, in 1938. (William T. Larkins) Bottom, VO-4 SOC-3 BuNo 1103 with white engine cowl and fuselage stripe. (USN)

Above, VS-9S SOC-1 BuNo 9896 at Grand Central Airport, Glendale, CA, in April 1936. Aircraft was coded 9-S-8, with the number 8 repeated in white under both wings. (William Swisher) Below, same VS-9S SOC-1 BuNo 9896 was assigned to the USS Salt Lake City. Tail stripe was white outlined in black. Vertical engine cowl stripe was also white outlined in black. (NAH)

Above, VS-9S SOC-1 BuNo 9894 at Grand Central Airport in 1936. Upper half of engine cowl was white. (John C. Mitchell via Swisher) Below, VS-9S SOC-1 rides the sled alongside the USS Salt Lake City (CA-25) on 2 April 1936. Tail stripe, fuselage stripe and engine cowl were white. (USN)

SCOUTING SQUADRON VS-10S

VS-10S SOCs assigned to the USS Louisville (CA-28) on 26 February 1937 were coded 10-S-13/14/15/16. #13 had the black cowl, #14 black top half of cowl, #15 black bottom half of cowl and #16 (bottom) had a black stripe at top of cowl. Tail stripe was yellow. (USN)

VS-10S was established on 1 October 1930 and the above insignia was submitted on 3 April 1935. It was later used by VCS-5 when VS-10S became VCS-5 on 1 July 1937. (SDAM via Tailhook)

Above, VS-10S SOC-1 BuNo 9925 assigned to the USS Chicago (CA-29) with lemon yellow tail stripe in 1936. Upper half of engine cowl was red. (via Swisher) Below, the USS Portland's (CA-33) SOCs in flight. They had black cowl markings. (USN)

SCOUTING SQUADRON VS-11S

SCOUTING SQUADRON VS-12S

Above, VS-11S SOC-1 BuNo 9902 while assigned to the USS Indianapolis (CA-35). Tail stripe was willow green and lower half of the engine cowl was white. (Raymond Pritchard via Larkins) Below, VS-12S SOC-1 BuNo 9913 coded 12-S-11 at Baltimore, MD, on 11 March 1937 was assigned to the USS New Orleans (CA-32). Tail stripe was black. (via Mark Aldrich)

VS-12S

Above, two VS-12S SOCs from beneath. Note the white number 5 under the left wing. (USN) Above, VS-12S SOC-1 after nosing-over on landing. Note that the horizontal black tail stripes were on the underside too. (USN) VS-12S SOC-1s from the USS Minneapolis (CA-36) on 2 November 1936 with black tail stripes. Cowl markings were white. (USN)

SCOUTING SQUADRON VS-14S

Below, VS-14S (14-S-1) SOC-1 BuNo 9921 assigned to the USS San Francisco (CA-38) in 1937. Fuselage stripe and cowl were red. The tail surfaces were aluminum as VS-14S did not have any tail colors assigned. VS-14S was only in service a couple of months before becoming VCS-8. (USN)

Bottom, VCS-2 (2-CS-8) SOC-2 BuNo 0391 at Portland, OR, while assigned to the USS Memphis (CL-13) in June 1940. Tail stripe was true blue and upper half of engine cowl was black. (Johnny Burnett via Larkins)

CRUISER SCOUTING SQUADRON TWO, VCS-2

CRUISER SCOUTING SQUADRON THREE, VCS-3

Above, VCS-3 SOC-1 BuNo 9856 assigned to the USS Cincinnati (CL-31) at Griffith Park National Guard Base, in 1938. Cowl and fuselage stripe was black. (William L. Swisher) Below, VCS-3 SOC-1 BuNo 9963 aboard the USS Cincinnati (CL-3). Tail stripes were red. (USN)

Above, two VCS-3 aircraft on their catapults. (USN) Below, VCS-3 SOC-3 assigned to the USS Philadelphia (CL-41) on 1 August 1940. Engine cowl and canopies were covered in protective canvas. (USN)

VCS-4

At right, VCS-4 SOC-1 BuNo 9866 4-CS-16 assigned to the USS Houston (CA-30). Note 4-CS-11 in the background which was assigned to the USS Pensacola (CA-24). (via Mark Aldrich) Below, VCS-4 SOC-1 BuNo 9961 assigned to the Houston on 17 August 1940. Tail stripe was blue. (USN)

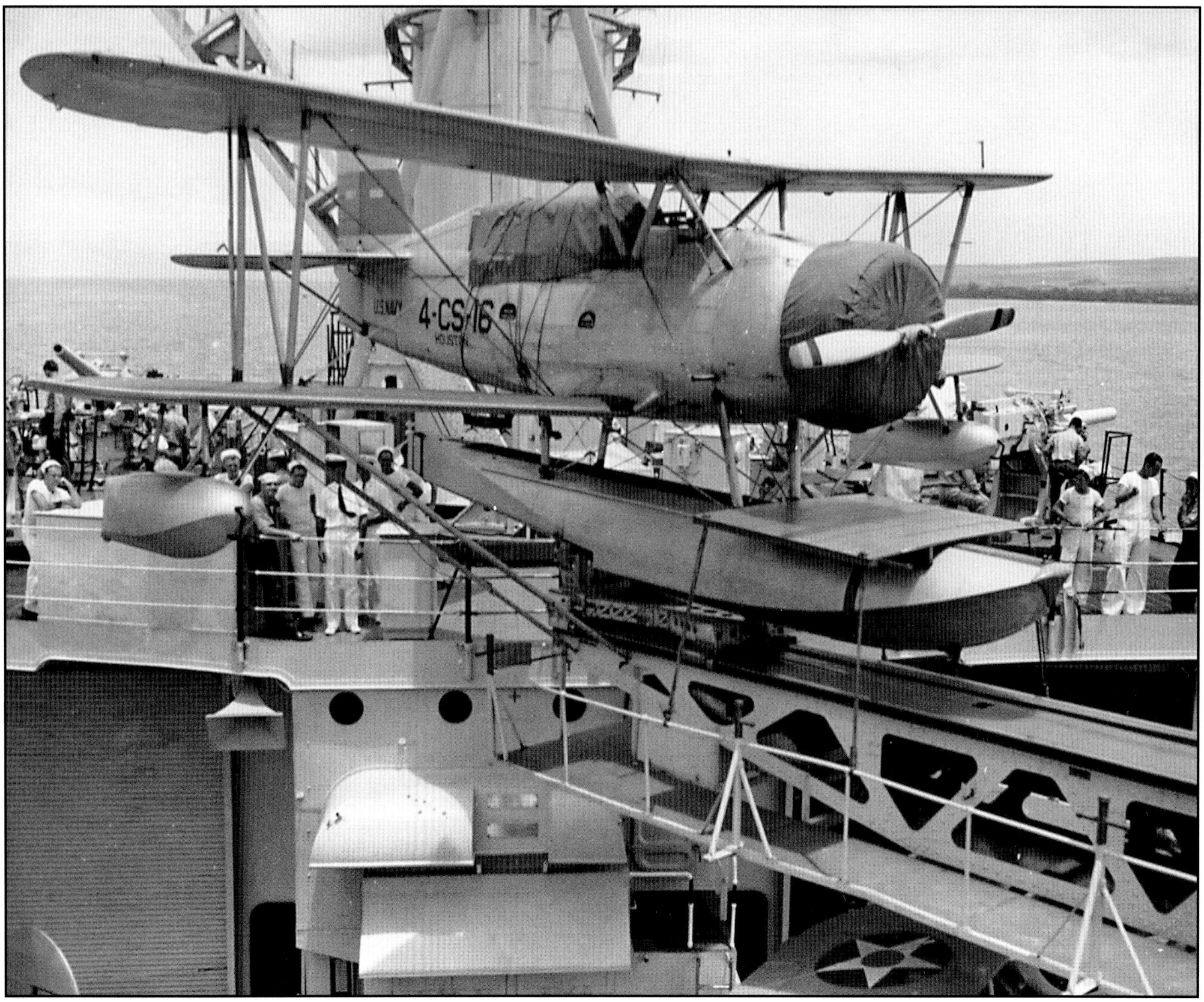

Above, VCS-4 SOC-2 4-CS-4 and SOC-1 4-CS-1 aboard the Northampton in August 1940. Tail stripe was blue. (USN) Below, VCS-4 SOC-2s aboard the USS Northampton (CA-26) at Pearl Harbor on 20 August 1940. Cowl markings were red. (USN)

At top right, VCS-5 SOC-2s assigned to the USS Chicago (CA-29) in July 1938. (USN) At right, SOC-1 BuNo 9866 in late 1937 while assigned to the Louisville. Engine cowl and fuselage stripe were black. (USN via Mark Aldrich) Bottom, VCS-5 SOC-1 BuNo 9866 taxiing at NAS Oakland, CA, was assigned to the USS Louisville (CA-28) in the fall of 1937. Tail stripe was lemon yellow outlined in black and the engine cowl and fuselage stripe were black. This aircraft later took part in the Arctic Operation Nanook in 1946, and Operation High-Jump in the Antarctic in 1947. (William T. Larkins)

CRUISER SCOUTING SQUADRON SIX, VCS-6

Above, VCS-6 SOC-1 BuNo 9915 on 28 January 1941. (USN) Below, three VCS-6 SOC-2s assigned to the USS Minneapolis (CA-36) on 3 January 1939. Tail stripe was black. Engine cowl markings were red. (USN)

VCS-6

At right, VCS-6 SOC-2 BuNo 0423 was assigned to the USS Minneapolis (CA-36) shown when landing at Oakland in June 1941. (William T. Larkins) Below, heavy cruiser four-aircraft detachment from VCS-6, SOCs 6-CS-5 white cowl and fuselage stripe, 6-CS-6 white top half cowl, 6-CS-7 white bottom half of cowl, and 6-CS-8 white vertical stripe on cowl. (USN)

VCS-6 SOC-1s in flight. (USN)

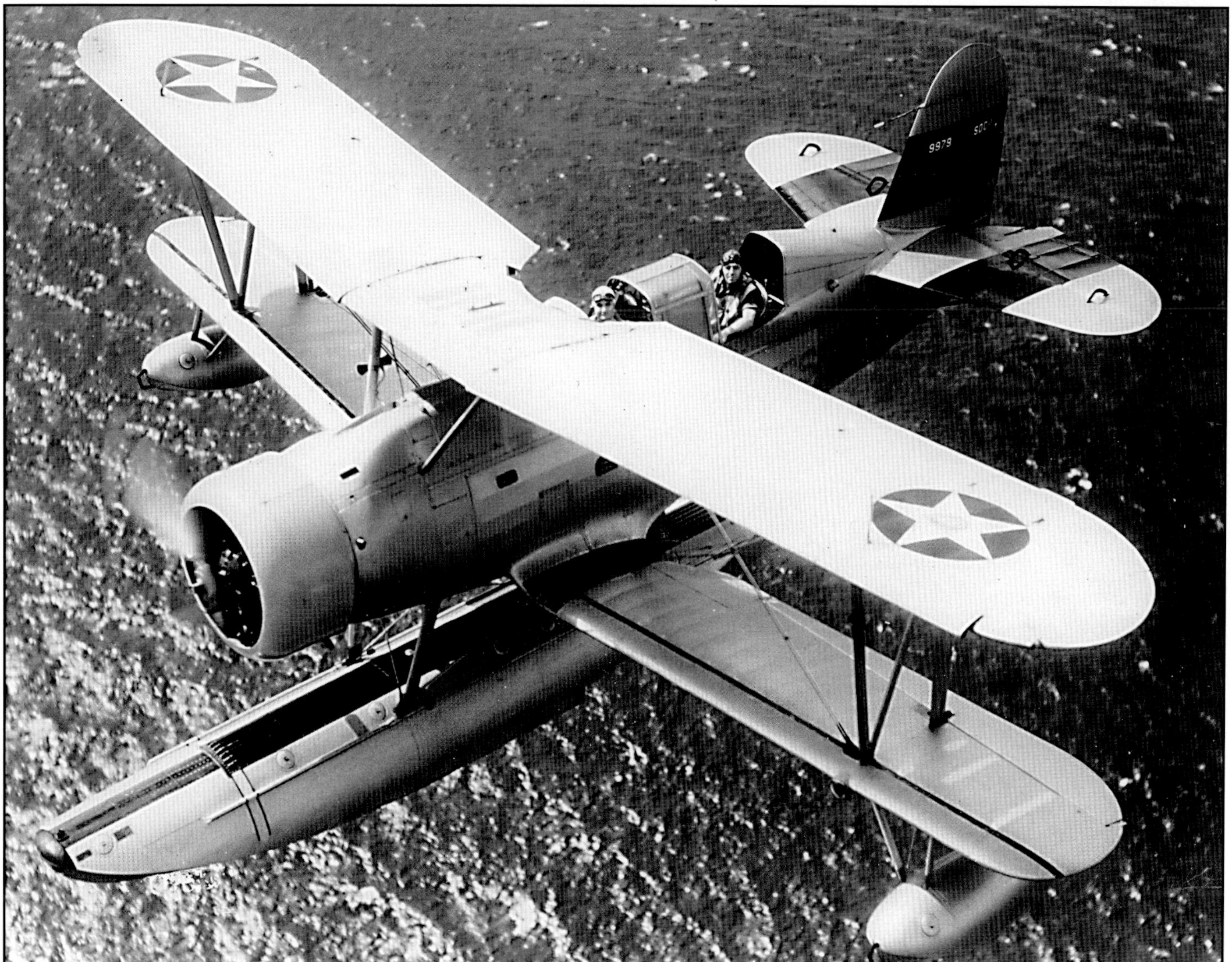

Above right and below, VCS-6 SOC-2 BuNo 0403 from the USS Minneapolis in 1938-1939. Note the squadron individual aircraft number "2" under the left wing tip. (USN)

Above and below, VCS-6 SOC-3s from the USS Minneapolis (CA-36) on 3 January 1939. (USN)

Above, VCS-6 SOC-1 BuNo 9908 assigned to the USS Astoria (CA-34) at Oakland in 1941. (Peter M. Bowers) Below, VCS-6 SOC on the well deck and hangar on the USS Minneapolis on 20 August 1940. (USN)

Above, VCS-6 SOC-1 BuNo 9935 assigned to the USS Astoria (CA-34) at the National Guard Base, Griffith Park, CA, on 2 April 1938. (William Swisher) Below, VCS-6 SOC-1 BuNo 9985 on the USS Minneapolis (CA-36) on 20 August 1940. (USN)

VCS-6

Above, VCS-6 SOC-1 BuNo 9870 assigned to the USS Astoria taxiing at NRAB Oakland in 1941. (Peter M. Bowers) At right, VCS-6 SOC-1 in July 1941. (Peter M. Bowers) Below, VCS-6 SOC-2 BuNo 8408 assigned to the USS Minneapolis (CA-36) at Oakland in 1941. (Frank Shertzer via Swisher)

CRUISER SCOUTING SQUADRON SEVEN, VCS-7

At right, VCS-7 SOC assigned to the USS Tuscaloosa (CA-37). (via Mark Aldrich) Below, VCS-7 SOC assigned to the USS Wichita (CA-45) in mid-1941 light grey overall paint scheme with all codes and lettering in white. (USN)

ABOVE, VCS-7 SOC assigned to the USS Wichita (CA-45) in mid 1941 light grey overall paint. (SDAM) Below, VCS-7 SOC-1 BuNo 9941 with red-bordered national insignia on 19 July 1943 while assigned to the USS Tuscaloosa (CA-37). Aircraft was written-off on 17 November 1943. (via Swisher)

VCS-8

Below, VCS-8 SOC-2 BuNo 0418 assigned to the USS Brooklyn (CL-40). Markings do not follow guidelines for the squadron or aircraft number 5 from the squadron. The engine cowl and fuselage stripe should be blue but are white in this photo. Vertical tail markings should be two thin gloss black stripes, but only one stripe is present and the tail appears to be painted a solid color, possibly red, beneath it. This may be due to the fact that VCS-8 was derived from the redesignation of VS-14S which had no official tail markings assigned, but may have used unofficial ones. (via Mark Aldrich) Bottom, VCS-8 SOC-3 BuNo 1143 assigned to the USS Savannah (CL-42) on 5 May 1938. Cowl and fuselage stripe were black; tail appears to be solid red, but should have two thin gloss black stripes instead. Fuselage stripe should not be on the turtledeck. (USN)

Above, VCS-8 SOC during Operation Torch (see pages 102-103). (USN) VCS-8 SOC assigned to the USS Philadelphia (CL-41) on 12 August 1940. (USN)

CRUISER SCOUTING SQUADRON NINE, VCS-9

At right, VCS-9 SOC-3s (9-CS-1, 9-CS-2, 9-CS-3, and 9-CS-4) from the USS Honolulu (CL-48) on 14 November 1940. (USN) Below, VCS-9 SON-1 BuNo 1175 assigned to the USS Honolulu (CL-48), seen at Oakland, CA, with double green tail stripes and red upper half of engine cowl. (via Peter M. Bowers)

Above, VCS-9 SOC-3s in flight from the USS Phoenix (CL-46). (USN) Below, VCS-9 SOC-3 from the USS Phoenix (CL-46). Closest aircraft is BuNo 1134. Cowl markings were blue; tail stripes were green. (NAH)

Above, VCS-9 SOC-3s parked at San Pedro, CA, on 8 November 1940. (USN) Below, VCS-9 SOC-3s from the USS Honolulu (CL-48) on 14 November 1940. (USN) Bottom, VCS-9 SOC-3 BuNo 1133 assigned to the USS Phoenix (CL-46) at NAS Oakland in November 1940 with double green tail stripes and blue lower half of engine cowl. (Peter M. Bowers)

Above, Helena-based SOC-3 coded 9-CS-13 at Oakland in 1941. (Peter M. Bowers via Swisher) At right, VCS-9 SOC-3 coded 9-CS-17 at Oakland in 1941 with wings folded. (William T. Larkins) Bottom, VCS-9 SOC-3 9-CS-14 at Oakland in 1941 allover light grey scheme while assigned to the USS Helena (CL-50). (Peter M. Bowers)

ESCORT SCOUTING SQUADRON 201, VS/VGS-201/VGS-1

The USS Long Island (AVG-1) was launched in January 1940 as the merchant cargo ship Monmacmail. The Navy acquired her in March 1941 and she was converted into the first prototype escort carrier in June 1941. Operational testing in the Atlantic resulted in a lengthened flight deck after which convoy escort and training carrier operations were conducted. The main aircraft type assigned to the Long Island were the SOC-3As from VS-201, later VGS-1. These aircraft were unique in that they carried a centerline depth bomb during their scouting/sub hunting missions. In May 1942, the carrier was transferred to the Pacific without the SOCs where it was utilized as an aircraft transport. It was reclassified ACV-1 in August 1942 and CVE-1 in July 1943.

Above, VGS-201 SOC-3As on the deck of the USS Long Island in May 1942. (NAH) Below, SOC-3A with depth bomb mounted on the centerline lands on the Long Island on 17 June 1942. (USN) Bottom, VGS-201 SOC-3A (1-GS-72) makes the 2,000th landing on the USS Long Island (AVG-1) on 20 April 1942. The aircraft was piloted by LCDR Lex L. Black, CO of VGS-201. Note lowered flaps and extended slats on the upper wing. (USN)

The USS Charger (AVG-6 later

CVE-30) was used for pilot training and anti-submarine patrols off the Atlantic coast and had SOC-3As from VGS-30 assigned.

Four other escort carriers besides the Long Island and the Charger operated SOCs. These were: the USS Bogue with VGS-9, USS Card with VGS-11, USS Core with VGS-13 and the USS Altamaha with VGS-18.

Above, VS-201 SOC-3A tied-down to the flight deck of the USS Long Island on 16 December 1941 with a centerline depth bomb mounted for anti-submarine operations. (USN) Below, VGS-30 SOC-3A on the deck of the USS Charger. (USN)

Above, VGS-30 SOC-3A conducts the first catapult launch from the USS Charger (AVG-6) on 5 November 1942. (USN) Below, VJ squadron SOC-3A after landing aboard the USS Charger (CVE-30) on 6 January 1944. Aircraft was painted in the tri-color scheme. (USN)

Above, CDR Monroe landing an SOC-3A aboard the USS Bogue (CVE-13) on 30 October 1942. (USN) Below, SOC-3A landing on the USS Ranger (CV-4) in 1942. (USN) Bottom, CDR J.J. Cassidy makes the first take-off from the USS Wasp (CV-7) in an SOC-2 in 1940. Note flotation gear installed on the forward fuselage and neutrality patrol markings. (NAH)

Two views of SOCs operating from the USS Ranger (CV-4) in 1942. (USN)

SOC OPERATIONS WWII

Among the first Navy aircraft to see action after the attack on Pearl Harbor were two SOCs from the USS Northampton (CA-26). At 1115 on 7 December 1941, LT M.C. Reeves in BuNo 9951 (5-CS-2) and ENS F.H. Covington in BuNo 9934 (5-CS-4) launched to search for the Japanese fleet 150-miles distant from the ship. Twenty minutes later while fifteen miles west of Kaui the SOCs were jumped by a Japanese A6M2 Zero. The Zero made seven passes on the SOCs putting fourteen holes in Reeves' aircraft and eleven into Covington's. The SOCs adopted a tight stepped-down formation and when the Zero was in range they turned to port to bring their rear guns to bear, allowing for visible hits on the Zero. The two gunners, ARM1c R. Payter and ARM2c J. Melton, expended some 175 rounds on the Zero and saw his aircraft smoking as it departed the area. Could this have been the Zero that crash-landed on the small island of Ni'ihau? Because of the SOC's low speed and tactics, the Zero was unable to properly lead the SOCs and press its attack home, a fact that was proved in prewar practice attacks made by Navy fighters against the Seagulls.

After Pearl Harbor, the SOCs fought their way across the Pacific conducting scouting operations, gunfire spotting, air crew rescue and an occasional bombing mission. In the Atlantic, they often carried depth bombs for use against German U-Boats.

In the Mediterranean, the SOCs saw heavy use during Operation

At right, SOC on the sled alongside the USS Augusta (CA-31) during Operation Torch off North Africa in November 1942. (NAH) At right bottom, Operation Torch SOC-3 from the USS Augusta (CA-131) on 4 December 1942. The observer is tossing a message to a surface ship. The large circle around the national insignia was yellow during the Operation. (USN) Below, two SOC-3s with depth charges under their right wings prepare for launch from a Brooklyn class heavy cruiser in January 1943. Raised deck area between the two catapults is the sliding hangar bay cover. (NAH)

At left, three VCS-5 SOC-3s in 1942. Censor has crudely removed the wing racks and bomb or depth charge loads with large white dots. (USN)

Torch and were marked with a yellow roundel surrounding the national insignia to help identify Allied aircraft involved in the joint British/American operation. They flew spotting missions over French Morocco in November 1942 and were attacked without damage by Vichy French Dewoitine D.520s fighters. They were even tasked with close air support and used 100 lb bombs and depth bombs on shore targets and a beached French submarine.

The SOCs then took part in the invasion of Sicily in July 1943. During Operation Husky an SOC from the Brooklyn was attacked by Bf 109s and was severely damaged but returned to the ship. Another SOC lost a crewman to friendly fire when he was blown out the aircraft and drowned. SOCs from the Boise and Savannah were also attacked by Messerschmitts. The lead SOC piloted by the Savannah's senior aviator, LT Charles Anderson, was shot down and crashed. Anderson was killed and his radioman Ed True survived. LTJG George Pinto landed nearby to give assistance, but his aircraft sank and he, his radioman and Ed True were rescued by the destroyer Ludlow. Two more Savannah SOCs launched with LT Osborn being shot down and the second SOC retiring to the Fleet for protection. The Boise also launched two more SOCs which were attacked by 109s too. Both aircraft returned safely to their ship. ENS Kalb Roher from Boise dropped bombs on a German tank column before being attacked by 109s. He too returned to the safety of the ship. On the subsequent mission this SOC was lost too. The Philadelphia would also lose three SOCs. The last SOC bombed and strafed italian positions and ended up herding 150 Italians to surrender by using its rear .30 cal. gun. With new SOCs, the Boise, Savannah, Brooklyn, and Philadelphia took part in the invasion of Italy.

Below, VCS-7 SOC-3s in outdoor storage at Lee-on-Solent airfield in Hampshire, England, in June 1944. During the invasion of Normandy the battleship and cruiser pilots switched to Spitfires. (NHC)

During Operation Overlord, the invasion of France, the SOCs were put ashore at Lee-on-Solent airfield in Hampshire and the pilots were given Spitfire Mk Vs for spotting purposes. The 17 Aviators were temporarily assigned to Air Spotting Pool #34 Reconnaissance Wing, Fleet Air Arm. They flew 209 sorties in Spitfires from 6 through 26 June 1944 with the loss of one Spitfire.

At right, SOC warming up on the catapult. Note rarely seen pilot's machine gun installation and war censorship of the underwing bomb positions. (USN) Below, late war photo of an SOC and a Fleet Boat operating close inshore probably on a combined pilot rescue mission. (USN)

At left and below, VCS-8 SOC being hoisted aboard the USS Philadelphia (CL-41) on 27 April 1942. (USN) Above, four SOCs on dollys on the afterdeck of an unknown ship in 1942. (USN)

After the attack on Pearl Harbor the SOCs were in the thick of things from the start of the US involvement in WWII. The Ford Island SOCs survived the attack and were immediately pressed into service patrolling the approaches to the Hawaiian Islands.

Above, anti-aircraft emplacements using .30 cal aircraft machine guns near the seaplane hangar on Ford Island in the days following the Pearl Harbor attack. Three SOCs and two PBYs can be seen in the background. (USN) Below, two SOCs and an OS2U stored under the trees on Ford Island in April 1942. (USN) At right, SOC in 1944. At right bottom, tri-color SOC being hoisted aboard an AV in 1946. (NHC)

Above, SOC on patrol off Newfoundland. Note unit markings have been censored from the fuselage side. (via Mark Aldrich)
Below, SOC-3 launching from an Astoria class heavy cruiser in late 1942. (USN)

Above, VCS-7 SOC at the end of its catapult stroke launches from the USS Tuscaloosa (CA-37) with a depth bomb under the left wing. (USN) Below, SOC being lowered onto one of the USS Tuscaloosa's (CA-37) catapults in August 1943. (USN)

Above, launching of an SOC from the USS Minneapolis (CA-36) on the morning of the Wake Island raid, 5 October 1943. (USN)
Below, attack on Wake Island by the USS Salt Lake City (CA-25) on 24 February 1942 with two SOCs on their catapults. (USN)
Above right, SOC launches from the USS Salt Lake City (CA-25) to spot during the shelling of Kiska Island on 2 August 1943. The Crew was LT R.P. Morse and ARM1C K.E. Waterman. The white stripe on the tail was used in the Aleutian area. (USN)
Below right, an SOC gunfire director is loaded on the catapult of the USS Minneapolis (CA-36) for launch on another Wake Island raid on 6 October 1943. (USN)

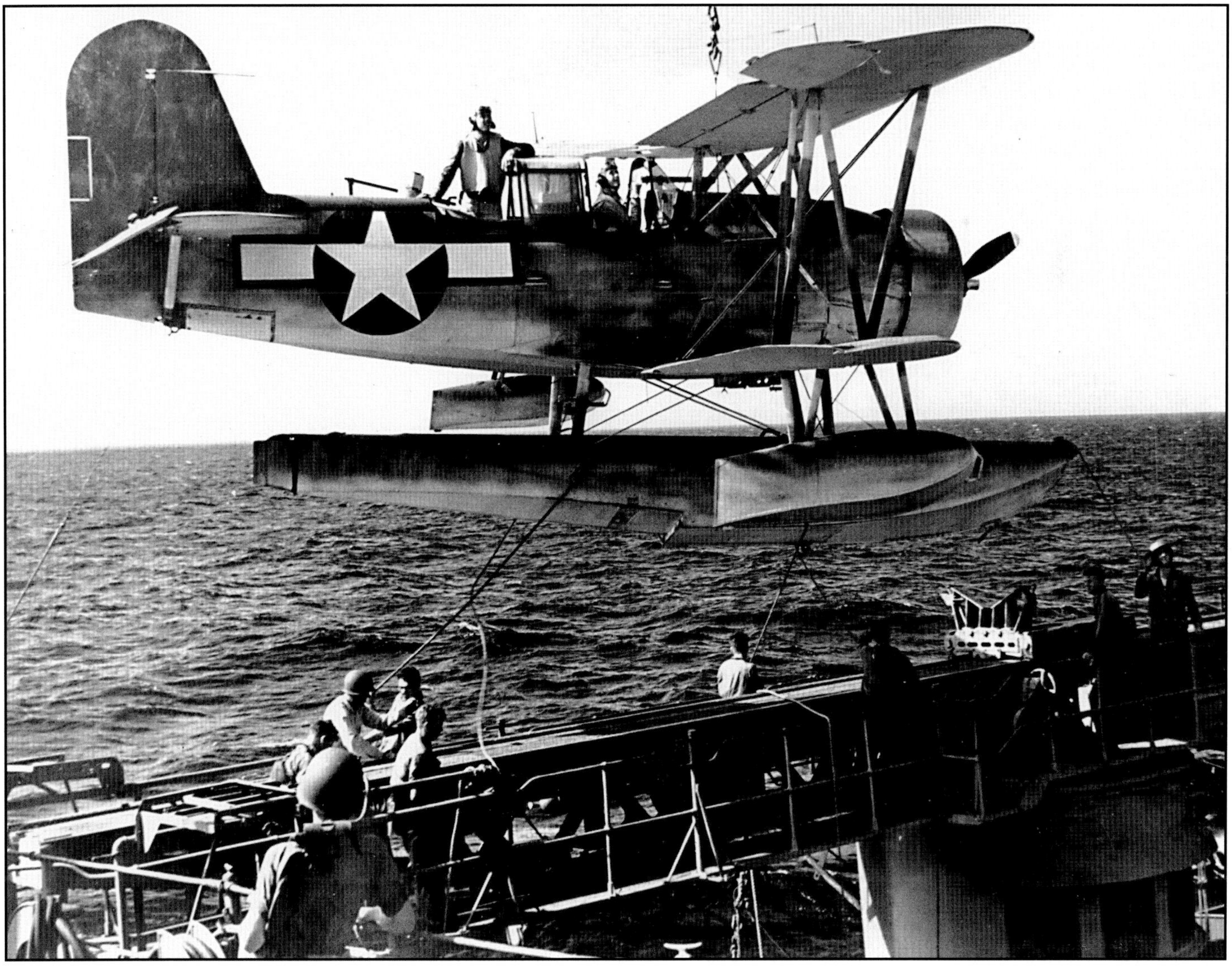

Above, SOC spots for gunfire for the USS Salt Lake City (CA-25), USS Northampton (CA-26), and for aircraft from the USS Enterprise (CV-6) over Wotje Atoll on 1 February 1942. (USN) Below, SOC-3 being hoisted aboard the USS Minneapolis (CA-36) after spotting for main battery firing on Kwajalein, Marshall Islands group, in January 1944. (USN) At right, SOC-3 leaving the catapult. (USN) Below right, SOC rests on the catapult of the USS Montpelier (CL-57) in 1944. (USN)

At left, SOC-1 pilot and radioman inspect damage to their upper wing caused by gunfire. Note damage to the lower wing too. (USN)

Above, unknown ship's Aviation Unit crew's photo. (Ginter collection) Below, SOC-3 on the USS Brooklyn (CL-40) being hoisted onto its catapult on 12 June 1944. There is a small number 9 on the engine cowl. (USN via Bob Rocker/Tailhook)

Above, SOCs aboard the USS Saint Louis (CL-49) in Tulagi Harbor on 5 July 1943. (USN) Below, two VCS-12 SOCs operating from the USS Denver (CL-58) on 14 February 1945. Note the white number 7 painted on the tail of the nearest aircraft. (via Dave Lucabaugh)

Flight and recovery sequence of an SOC from the seaplane tender USS Curituck (AV-7) in January 1947. (USN)

Tri-color SOC on the stern of an AV in 1946. Note tail of the PBM behind it. (NHC)

SOSU-3 SCOUT OBSERVATION SERVICE UNIT

Above, Scout Observation Service Unit (SOSU-3) SOC-2 at NAS Alameda, CA, had a very unique cheat line on the engine cowl between the upper fuselage sea blue and undersides NS insignia white colors. (Fred Dickey) At right and below, gloss sea blue SOSU-3 SOC-2 BuNo 0402 in late 1946 with two SC-1s in the background. The large fuselage numbers W60 were painted orange/yellow. (Peter M. Bowers)

Above, tri-color SOC over the Fleet in 1945. (USN) Below, one of the last SOCs in inventory, overall gloss sea blue SOC-3 on 19 April 1947. (SDAM)

SELECTED SHIPS THAT OPERATED SOCs

Above, USS Long Island (AVG-1) on 11 November 1941 with its SOC-3As on deck. (USN)

At right, USS Montpelier (CL-57) at Efate, New Herbites, on 22 April 1942. Note oversized national insignia on the SOC in the foreground. (USN)

Below right, USS Patoka (AO-9) at Balboa, Canal Zone, on 22 February 1940. (USN)

Below, USS Tennessee (BB-43) with pre-war SOCs atop turret and on fantail catapults. (via Mark Aldrich)

HASEGAWA 1/72 SCALE CURTISS SOC-3 SEAGULL

In the 1970s, the 1/72nd modeler was treated to an SOC-3 Scout Observation kit produced by Hasegawa. It was available in the United States in both the smaller Hasegawa box (kit number JS-057) and a larger re-box offered by Mini Craft Models. The kit was a bit crude by Hasegawa's current standards which might explain why we have not seen a re-issue of it since then.

The kit offered either wheels or float options, and choice of pre-war markings. It contained forty-six parts. Examples below were one-day builds from the 70s with no extras. Lack of rigging really detracts from the finished product.

A comprehensive decal sheet was offered by Microscale (72-62) which offered VCS and VO squadrons, Coast Guard, USMC (VMS-2), and utility units including the USS Saratoga (CV-3).

1/48 SCALE VAC WINGS 48 SOC VACUFORM KIT

vac Wings 48™

Exquisite vac-forms for the connoisseur modeler.

The Vac Wings 48 SOC Seagull offering is a mixed media kit. It comes in three vacuform sheets and a bag of white metal parts which contains: engine, landing gear, wing struts and propeller. My copy also included a decal sheet with pre-war national insignia.

Yellow Wings Decals provides two aftermarket decal sheets for the SOC. One includes markings for VO-1 and VO-2 assigned to USS Arizona, USS Nevada, USS Pennsylvania, USS California, USS Tennessee, and USS Oklahoma. The other for VO-3 and VO-4 on USS Idaho, USS Mississippi, USS New Mexico, USS West Virginia, USS Colorado, and USS Maryland.

Due to time constraints the kit shown here was a quick build and is incomplete. Aileron struts, rigging, and detail parts like pitot tube, elevator hinges, and bomb racks have not been fabricated.

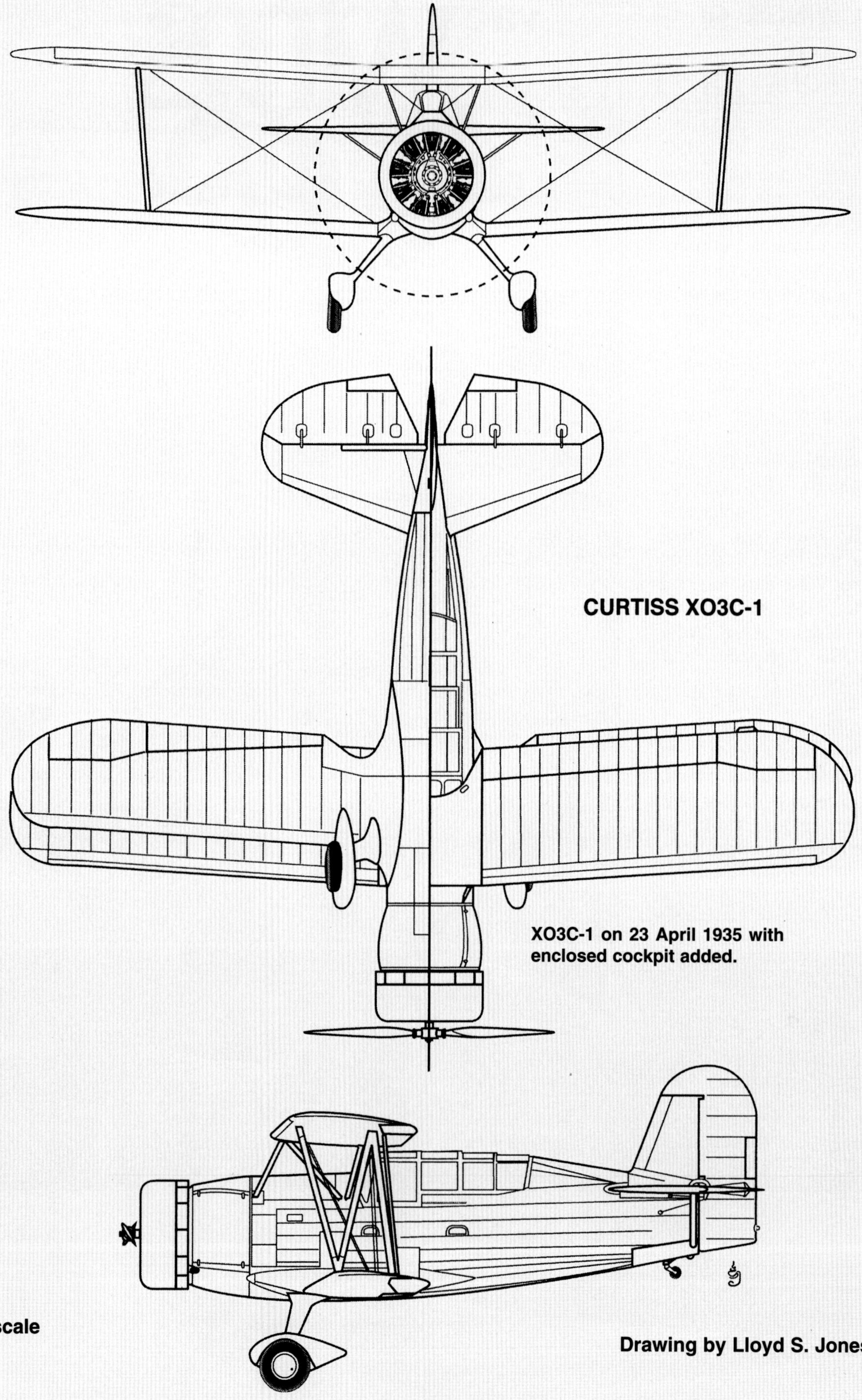

CURTISS XO3C-1

XO3C-1 on 23 April 1935 with enclosed cockpit added.

1/72 scale

Drawing by Lloyd S. Jones

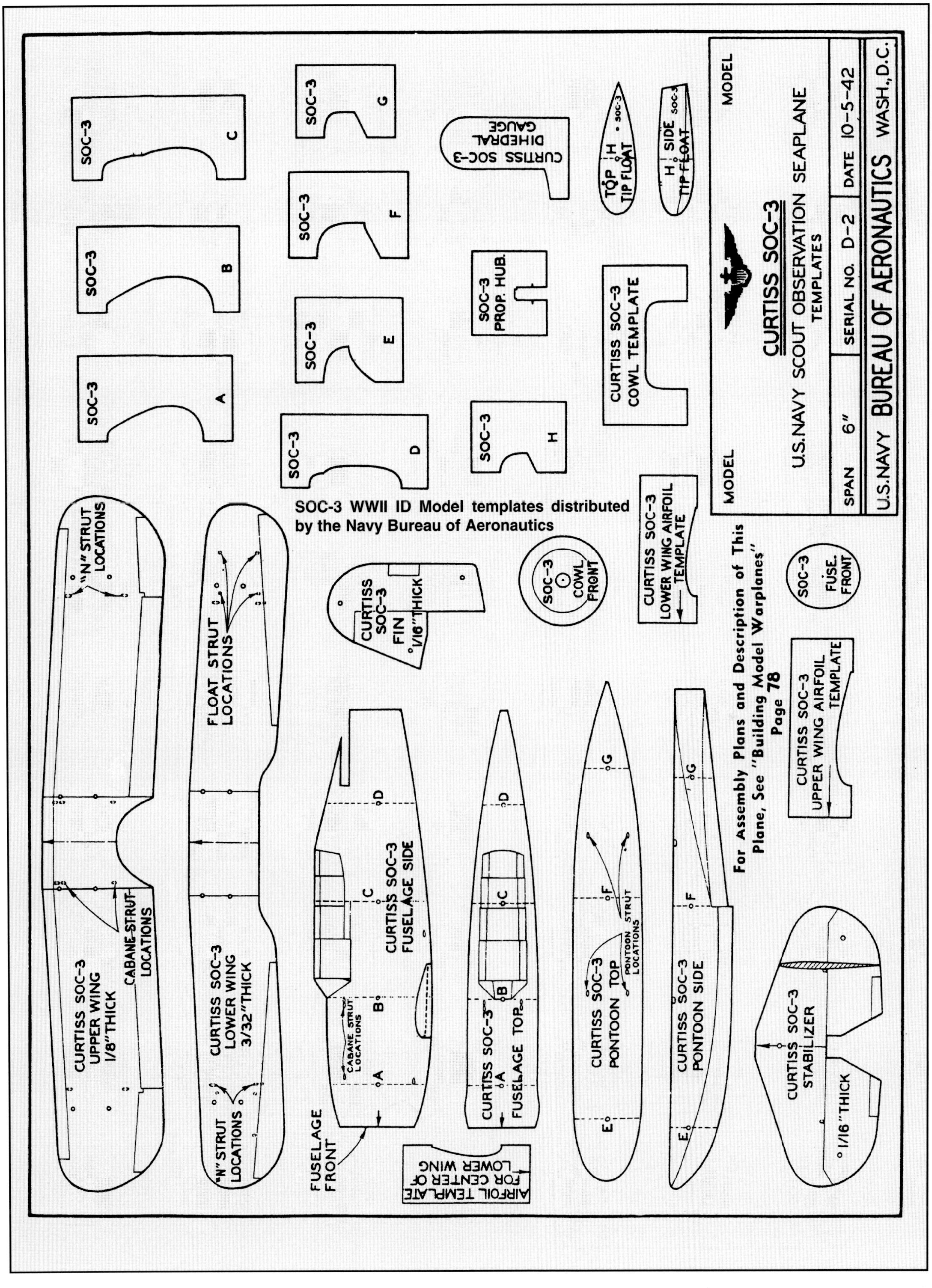

SOC-3 WWII ID Model templates distributed by the Navy Bureau of Aeronautics

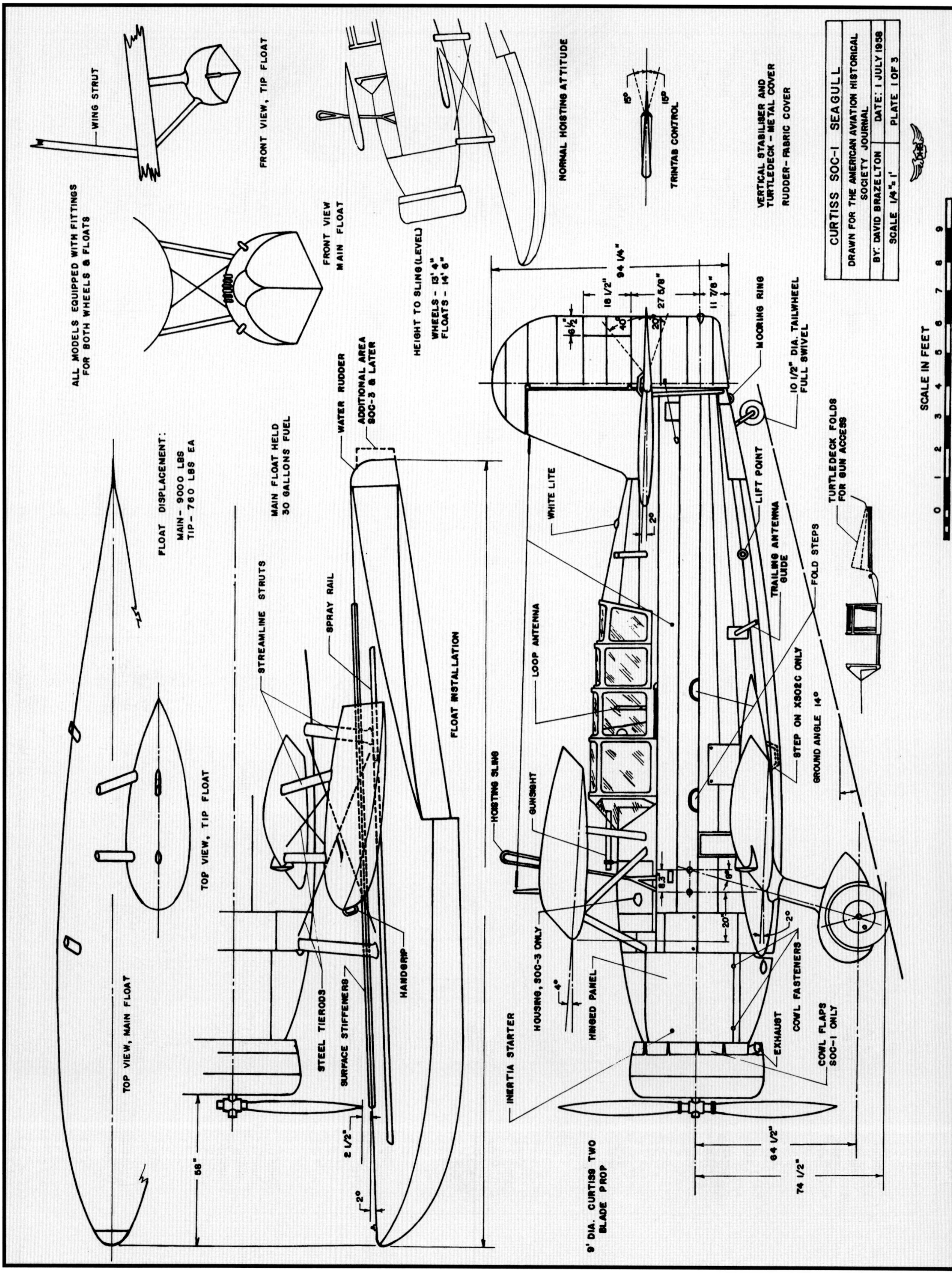
CURTISS SOC-1 SEAGULL
DRAWN FOR THE AMERICAN AVIATION HISTORICAL SOCIETY JOURNAL
BY: DAVID BRAZELTON
DATE: 1 JULY 1958
SCALE 1/4"=1'
PLATE 1 OF 3
SCALE IN FEET
0 1 2 3 4 5 6 7 8 9
ALL MODELS EQUIPPED WITH FITTINGS FOR BOTH WHEELS & FLOATS
WING STRUT
FRONT VIEW, TIP FLOAT
FRONT VIEW MAIN FLOAT
NORMAL HOISTING ATTITUDE
TRIMTAB CONTROL
15°
15°
VERTICAL STABILISER AND TURTLEDECK - METAL COVER
RUDDER - FABRIC COVER
HEIGHT TO SLING (LEVEL)
WHEELS - 13' 4"
FLOATS - 14' 6"
FLOAT DISPLACEMENT:
MAIN - 9000 LBS
TIP - 760 LBS EA
MAIN FLOAT HELD 30 GALLONS FUEL
WATER RUDDER
ADDITIONAL AREA SOC-3 & LATER
94 1/4"
16 1/2"
27 5/8"
11 7/8"
6 1/2"
40°
20°
2°
MOORING RING
10 1/2" DIA. TAILWHEEL FULL SWIVEL
WHITE LITE
LIFT POINT
TRAILING ANTENNA GUIDE
FOLD STEPS
TURTLEDECK FOLDS FOR GUN ACCESS
STREAMLINE STRUTS
SPRAY RAIL
FLOAT INSTALLATION
LOOP ANTENNA
STEP ON XSO2C ONLY
GROUND ANGLE 14°
TOP VIEW, TIP FLOAT
TOP VIEW, MAIN FLOAT
HOISTING SLING
GUNSIGHT
HOUSING, SOC-3 ONLY
HINGED PANEL
4°
INERTIA STARTER
COWL FASTENERS
EXHAUST
COWL FLAPS SOC-1 ONLY
STEEL TIERODS
SURFACE STIFFENERS
HANDGRIP
2 1/2"
2°
58"
9' DIA. CURTISS TWO BLADE PROP
64 1/2"
74 1/2"